HIGHLY EFFICIENT METHODS FOR SULFUR VULCANIZATION TECHNIQUES, RESULTS AND IMPLICATIONS

SELECTION AND MANAGEMENT OF RUBBER CURATIVES

DR ALI ANSARIFAR

BALBOA.PRESS

A DIVISION OF HAY HOUSE

Balboa Press books may be ordered through booksellers or by contacting:

Balboa Press
A Division of Hay House
1663 Liberty Drive
Bloomington, IN 47403
www.balboapress.co.uk
UK TFN: 0800 0148647 (Toll Free inside the UK)
UK Local: (02) 0369 56325 (+44 20 3695 6325 from outside the UK)

Because of the dynamic nature of the Internet, any web addresses or links contained in this book may have changed since publication and may no longer be valid. The views expressed in this work are solely those of the author and do not necessarily reflect the views of the publisher, and the publisher hereby disclaims any responsibility for them.

The author of this book does not dispense medical advice or prescribe the use of any technique as a form of treatment for physical, emotional, or medical problems without the advice of a physician, either directly or indirectly. The intent of the author is only to offer information of a general nature to help you in your quest for emotional and spiritual well-being. In the event you use any of the information in this book for yourself, which is your constitutional right, the author and the publisher assume no responsibility for your actions.

Any people depicted in stock imagery provided by Getty Images are models, and such images are being used for illustrative purposes only. Certain stock imagery © Getty Images.

Print information available on the last page.

ISBN: 978-1-9822-8504-3 (sc)
ISBN: 978-1-9822-8506-7 (hc)
ISBN: 978-1-9822-8505-0 (e)

Balboa Press rev. date: 06/21/2022

CONTENTS

The rubber industry is a success story in the manufacturing endeavours of mankind. It started with a tree that was grown in tropical South America, and then the rubber production spread to other tropical countries in different parts of the world. The rapid expansion of the rubber industry and the range of products that it produces are truly astonishing. Rubber is a fascinating material to study because of its unusual properties. It is even more amazing that so many products for so many applications are made using rubber. The success of the rubber industry is due to the availability of a range of chemicals to cure or vulcanize rubber for industrial applications. These chemicals are toxic, and as a result, there are laws and regulations which limit their use. This poses a major challenge to the manufacturers and users of rubber chemicals, and urgent steps must be taken to address this issue. This can only be achieved when the excessive use of chemical curatives in rubber vulcanization is reduced. In this book, two new methods are described and supported by extensive experimental results that address this problem and offer a way forward. A software program was developed to help with the selection and management of some selected chemicals for sulfur vulcanization. It is time to make rubber compounds as green as the rubber tree. Green rubber products are now within reach.

PREFACE

Sulfur vulcanization, also called curing or crosslinking, is a major stage in the processing of raw rubber into a practical industrial product. To aid the reaction of sulfur with rubber at high temperatures, some chemicals are added to control the onset, rate, and extent of the reaction of sulfur with the rubber. These chemicals are called accelerators and activators. The use of accelerators and activators in sulfur vulcanization has been so successful that in most rubber formulations, there are now two accelerators (primary and secondary) and two activators (primary and secondary). It is true that sulfur vulcanization is a more efficient process today than it was at the time of Charles Goodyear in 1844, but serious health and safety issues related to the excessive use of these chemicals in rubber have emerged. These chemicals are damaging to aquatic life and the environment, and so, their use is restricted by various laws for the environment, health, and safety. For example, by the European Directives 2004/73/EC and 67/548/EEC. Furthermore, the exact amounts of the chemicals required for curing rubber are not measured, and there is no reason why so many chemicals are used in industrial rubber compounds today.

I am a materials scientist and have worked with rubber for the last 32 years of my professional life. I have published around 140 technical research papers in peer-reviewed international scientific journals and in technical magazines for the polymer and tire industries. I have contributed chapters to scientific books and carried out consultancy work for some companies and manufacturers in the UK. The most varied aspects of my experience were with the suppliers and users of rubber chemicals. I searched some rubber formulary literature for industrial rubber products and learned about the toxic properties of the chemicals used in the compounding of rubber published by the European Chemicals Policy, Registration, Evaluation, Authorization, and Restriction of Chemicals (REACH). I concluded that the current methods for using chemicals in sulfur vulcanization are fundamentally flawed. So, I initiated a new study and supervised research projects to measure the chemical curatives for sulfur vulcanization more accurately.

This book is a collection of extensive experimental work and measurements that were carried out in several postgraduate and doctoral projects supervised by the author and two colleagues, Dr. George W. Weaver and Professor K. G. Upul Wijayantha, in the Department of Chemistry at Loughborough University, U.K. The aim is to promote a better way of using the chemical curatives in sulfur vulcanization and simplify the compounding of rubber with the curatives.

Two highly efficient methods for the sulfur vulcanization of unsaturated hydrocarbon rubbers have been developed. One method measures the exact amount of a sulfenamide accelerator to cause a reaction between sulfur and rubber to form chemical crosslinks and then adds zinc oxide to improve the efficiency of the curing process. Another method of treating the surface of zinc oxide with a sulfenamide accelerator and a thiuram accelerator is by evaporating a suspension of zinc oxide in a solution of the accelerators in an organic solvent to provide two convenient single material components to use as additives. The exact amounts of each single additive for curing natural rubber (NR), polybutadiene rubber (BR), and ethylene-propylene-diene (EPDM) rubber were measured. Both methods eliminated the need for the secondary accelerator, secondary activator, and stearic acid from curing, and only the optimum amount of each additive was used to vulcanize the rubber. As a result, the rubber was cured with fewer chemicals. This was highly beneficial to health, safety, and the environment, and made compounding rubber with the chemical curatives a lot cheaper and safer. A software program was developed to help with the selection and management of the zinc oxide and the two accelerators for sulfur vulcanization.

It is hoped that the producers and users of chemicals in the rubber industry will benefit from this work. The two additives are highly efficient in curing rubber and will help to address the concerns expressed above. The compounding of rubber with fewer chemicals will be much safer, more environmentally friendly, and very cost-effective. A limited use of chemicals in vulcanization will advance what Charles Goodyear started some years ago and improve the environmental credentials of the rubber manufacturing industry world-wide. The rubber industry must improve the safety of its products to meet its obligations under various laws and regulations. It is time to solve some of these pressing issues and make the rubber industry green.

CHAPTER 1

1 Natural rubber and synthetic rubbers - A historical perspective

Natural rubber (NR) is extracted from the latex of Hevea brasiliensis, a tree indigenous to South America that has been cultivated for many years in the Far East and in Africa. The coagulum, known as "Field Latex," is obtained by tapping the tree. Tapping involves introducing a cut into the skin of the tree and collecting an aqueous-based serum or mixture in a small cup filled with some ammonia to preserve it against bacterial attack. The latex contains a high proportion of a hydrocarbon, mixed with proteins, resins, and other constituents. The analysis has shown the latex to have: 94.5 % hydrocarbon; 2.8% acetone solubles; 0.4 % nitrogen; and 0.2 % ash. The structure of the hydrocarbon is essentially 1,4-polyisoprenoid, a polymer of very high molecular weight [1.1]. In many applications, NR acts as a spring and carries a high load under compression, yet functions at high strains and low stiffness compared with other materials such as metals. NR can store more elastic energy per unit volume than steel because it possesses some inherent damping properties. Moreover, it can deform elastically by several hundred percent without failing. NR can extend a lot under modes of deformation which do not constrain it hydrostatically, but much less so when, for example, hydrostatic compression is involved. The unusual elastic properties of the rubber make it suitable in diverse engineering applications where both static and dynamic conditions are present [1.2,1.3]. Since these unique properties were discovered, NR has been the material of choice for engineers. For years, rubber chemists have strived to make new polymers by emulating the unique properties of NR. There are now a lot of synthetic rubbers (SRs) on the market that are used to manufacture a wide range of industrial products. The rubber industry is a world-wide supplier and manufacturer for a range of industries such as transportation, communication, construction, aerospace, energy, marine, education, and the healthcare sector [1.3]. A brief review of the structure, composition, properties, and applications of NR and SRs will suffice here.

1.1 Crystallization in natural rubber

NR and other polymers have long molecules. Parts of the long molecules that have a regular structure arrange themselves into a regular state depending on the ambient temperature and the extent of deformation. This is when the material exhibits a first-order transition and becomes denser, resembling the crystallization of simple liquids. Because of molecular imperfections and chain entanglements, the degree of crystallinity does not exceed about 30% in NR [1.4]. This limited degree of crystallization has a profound effect on the properties of the rubber. The material becomes harder and loses its extensibility and elasticity significantly. When a sample of rubber is held in tension at a fixed strain, the tensile force decays and may approach zero as crystallization advances. When the tensile force is removed from the sample, little or no recovery takes place. Crystallization in NR is affected by both temperature and the strain on the rubber. When a sample of rubber is heated above its normal crystalline melting temperature, it becomes soft and regains its elasticity. It can also crystallize under the application of strain. The crystalline melting temperature for NR depends on the crystallization temperature. For example, if the rubber is crystallized at -40°C, it starts melting at -34°C, and the melting completes at -2.5°C. If it is crystallized at 15°C, it starts melting at 22°C, and the melting completes at 32°C [1.5]. However, if the sample is cooled so rapidly that the molecular chains become immobile before they can rearrange themselves into a regular structure, the material will be amorphous at a temperature below its freezing point. For optimum conditions for crystallization, the temperature and the deformation are such that the molecules have enough energy and freedom to form an ordered structure but not too much to dissociate under the influence of random thermal motion. Crystallization under normal conditions is not desirable because it hardens the raw rubber in storage and causes loss of elasticity. This is highly detrimental to engineering components such as vibration management systems (VMS), like engine mountings, and can severely damage and render the VMS unserviceable. At large deformations, a high degree of crystallinity in the material (termed "strain-induced" crystallization) improves strength and increases tear resistance [1.5]. Strain-induced crystallization is a highly desirable property, and rubber scientists and technologists have been trying for years to produce polymers emulating crystallization in NR.

1.2 Structure, composition, properties, and applications of rubbers

1.2.1 Natural rubber (NR) and synthetic polyisoprene (IR)

As mentioned before, NR is a crystallizing polymer containing over 99% of linear cis-1,4 polyisoprene. The average molecular weight of the polyisoprene in NR ranges from 200,000 to 400,000, and the rubber has a broad molecular weight distribution. The rubber has 3000 to 5000 isoprene units per polymer chain, and each isoprene unit has one double bond. The double bonds in the chains make the rubber an unsaturated hydrocarbon polymer. The double bonds are essential for the reaction of sulfur with the rubber chains to produce stable covalent crosslinks, but they can also react with oxygen or ozone. This is termed oxidative and ozone ageing, respectively, and is highly damaging to the rubber properties and shortens the service life of rubber components. Other reactions can occur with hydrogen, chlorine, or hydrogen chloride in the presence of the double bonds, which produce new polymers from NR. If NR is cooled very slowly from 10°C to -35°C, it becomes non-transparent and hard, and loses its elasticity due to partial crystallization. When NR is stretched more than 80% of its original length, strain-induced crystallization occurs as the polymer chains orient along the direction of the applied load. The crystallization improves the intermolecular attractive forces, and this reinforces the strength of the rubber along the axis of deformation and causes a lower tensile strength perpendicular to it. This is termed "anisotropy." The rubber is non-polar because it is halogens-free in its structure. The temperature at which the rubbery state changes into a glassy one is termed the "glass transition temperature" or T_g. NR has a T_g of about -71°C. The rubber chains are thermally mobile at room temperature, but as the rubber is cooled slowly and progressively, it becomes harder until it finally loses its elasticity and becomes a solid. NR has high gum strength (gum is unfilled raw rubber), high tear strength, long flex life, high compression set resistance at ambient temperatures, good low temperature flexibility in the absence of crystallization under slow progressive cooling, and high abrasion resistance. NR has some limiting properties too. For example, moderate to poor resistance to heat, moderate to poor resistance to non-polar oil and solvents, moderate to poor resistance to oxygen, ozone and ultraviolet light (UV). It is a highly inflammable material and burns easily and has a working temperature range of -55°C to

100°C. NR is used to produce tires, hoses, belts, all kinds of vibration management systems, bridge bearings, gloves, and balloons [1.1,1.6]. All in all, NR is a highly versatile polymer that is in use in many applications. The synthetic analogue of NR is chemically and structurally very similar to it in that it contains olefinic double bonds and hence can be vulcanized with sulfur. It is referred to as synthetic NR or IR. Some grades of IR have up to 98% cis-1,4 content, which is slightly lower than that of NR. The lower cis- 1,4 content in IR adversely affects the self-reinforcement due to strain crystallization and results in an inferior green strength of compounds and lower tensile properties for the cured rubber [1.7].

1.2.2 Polybutadiene rubber (BR)

Polybutadiene (BR) is a non-polar unsaturated hydrocarbon polymer made of butadiene units which are joined linearly by 1.4 as well as 1.2-addition. The T_g of BR ranges from -75°C to -100°C, depending on the cis-1.4 content in the molecular chains [1.8]. The commercial grades have a cis-1.4 content of about 96%. The rubber has the best abrasion resistance when it is made of pure cis-1.4 structure. But when the 1.2 content increases, the abrasion resistance deteriorates. The properties of NR, or styrene-butadiene rubber (SBR), are improved through blending with cis-1.4 BR to produce polymers with low T_g, which possess high abrasion resistance, and high resilience. The blending of BR with NR and SBR also improves the heat built-up, the rolling resistance and the resistance to cracking in tires. A high volume of BR is used in the manufacture of tires, shoe soles, bumpers, transmission belts, roll covers, conveyor belts, and shock absorber pads [1.7].

1.2.3 Styrene-butadiene rubber (SBR)

Styrene-butadiene rubber (SBR) is a non-polar unsaturated hydrocarbon polymer and cannot crystallize because of the irregular arrangement of the monomers in its molecular chains. SBR is made of two dissimilar monomers where the ratio of butadiene to styrene is mostly 76.5 wt% to 23.5 wt%. As the styrene content in the SBR increases, the T_g rises and resilience decreases. For example, for SBR rubber with 23 wt% styrene content, a T_g of -60°C has been reported [1.7, 1.8]. Normally, rubbers with very low T_g values, possess high resilience and good abrasion resistance, whereas those rubbers with high styrene content and correspondingly high T_g, show low resilience and poor abrasion resistance. The SBR vulcanizates have by far better dynamic fatigue resistance, ageing resistance, and heat resistance than the NR vulcanizates, and provided they are compounded with reinforcing solid fillers such as carbon black, they are superior in performance to comparable ones from NR. SBR, in combination with BR and NR, is used mainly to produce car and light truck tires. Other applications of SBR include belting, cable insulation, hoses, shoe soles, surgical equipment, food packaging, roll coverings, and pharmaceuticals [1.7].

1.2.4 Ethylene-propylene-diene rubber (EPDM)

Ethylene-propylene-diene rubber (EPDM) is a non-polar unsaturated hydrocarbon polymer. It is non-crystallizing because of its irregular chain structure and hence has poor gum strength. It is formed when the crystallization of polyethylene at room temperature is inhibited by incorporating propylene segments into its chain structure. This produces an amorphous rubbery polymer (EPM) with a T_g below room temperature that cannot be cured with sulfur because it is saturated (no double bonds in its structure). So, during the production of

EPM, a diene monomer is added to produce an unsaturated polymer that can be cured with sulfur. These rubbers are called EPDMs. In commercial EPDMs, only three dienes are used, and the double bonds reside in the side-groups of the polymer chains. The dienes are dicyclopentadiene (DCP), ethylidene norbornene (ENB), and trans-1.4 hexadiene (HX). For EPDM rubbers, T_g values ranging from -40°C to -60°C have been reported [1.7, 1.8]. The EPDMs have excellent resistance to ageing by oxygen, ozone, water, heat, and weathering, and are used in a working temperature range of -50°C to 150°C. EPDMs are commonly used in gaskets, conveyor belts, steam hoses, cable insulation, roofing sheets, sidewall compounds of tires, automotive products such as radiator hoses and windows, plastic modifiers, and dock fenders [1.7].

1.2.5 Acrylonitrile-butadiene rubber (NBR)

Acrylonitrile-butadiene rubber (NBR) is a polar unsaturated hydrocarbon polymer with an irregular chain structure, and as a result, it cannot crystallize on extension and possesses poor tensile properties. The acrylonitrile content ranges from 18% to 51% by weight, and as the acrylonitrile content increases, the polymer becomes more polar and the swell resistance to non-polar oils and solvents improves. Moreover, when the acrylonitrile content increases, the T_g rises, the elasticity and low temperature flexibility deteriorate, but the processability of the compounds improves. The T_g of an NBR with an acrylonitrile content of 20 wt% is about -35°C and that of an NBR with 40 wt% acrylonitrile content is about -20°C [1.7, 1.8]. The working temperature range for this rubber is -20°C to 120°C. NBR is used in applications where there is a requirement for good resistance to swelling in oils and gasoline, and good resistance to heat ageing and abrasion. For example, in O-rings and valves, static seals, hoses, roll coverings, conveyor belts, linings, printing blankets, work boots, impact modifiers in plastics, and in products for the food industry. NBR has excellent bonding properties because of its polarity [1.7].

1.2.6 Butyl rubber (IIR)

Isoprene-isobutylene rubber, or butyl rubber, is a non-polar unsaturated hydrocarbon polymer consisting of 97 to 99.5 mole% isobutylene and 0.5 to 3 mole% isoprene. The isoprene monomer is unsaturated and provides the double bond needed for sulfur vulcanization. The butyl chain is largely saturated because of the high isobutylene content. Hence, the polymer has good resistance to ageing by oxygen and ozone, and low gas permeability, but the vulcanization rate is slow because of low levels of unsaturation. When compared with SBR and NBR, unfilled or gum IIR vulcanizates have higher tensile strengths, which can be improved further with reinforcing fillers. The heat resistance of sulfur vulcanized IIR is inferior to that of other vulcanizates, such as sulfur-cured EPDM. IIR is used in cable insulation and jacketing, curing bladders, roofing membranes, pharmaceutical stoppers, and inner tubes of tires [1.7].

1.2.7 Polychloroprene rubber (CR)

Poly-2-chlorobutadiene, or chloroprene rubber (CR), is a polar unsaturated hydrocarbon polymer. It contains one chlorine atom for every four carbon atoms in the chain of polymers. There are three grades of the polymer, low, medium, and high crystallizability, depending on the regularity of the chain structure. The formation of more regular chain structures increases the rate of crystallization of the polymer. The grades that crystallize at high rates, harden very rapidly, and lose elasticity, are thus less suitable to produce rubber goods. To retard the tendency to crystallize, small amounts of styrene or acrylonitrile monomers are added to increase the irregularity of the polymer chains. CR has good resistance to swelling in mineral, animal, and vegetable oils and fats. It also has better flame, weather, and ozone resistance than other unsaturated hydrocarbon rubbers because of the chlorine atoms in its structure. Unfilled or gum vulcanizates have a higher mechanical strength than those from most types of SRs because of the ability of CR vulcanizates to strain-crystallize. CR vulcanizates have very good flame resistance and self-extinguish quickly after the removal of the flame. Some grades of CR with a regular chain structure harden on cooling. This is caused by crystallization and adversely affects the low temperature flexibility of the rubber. CRs are used in rolls, belts, seals, hoses, bearings, linings, rubberized fabrics, extrusions, window and construction profiles, automotive applications, and in the production of contact adhesives [1.7].

1.3 Sulfur vulcanization

1.3.1 Sulfur

There are many chemicals available to vulcanize rubber. The most commonly used vulcanizing agents are sulfur, sulfur donors, thiuram disulphides, peroxides, metal oxides, and resins. The most important chemical for curing rubber is sulfur. Sulfur with at least 99.5% purity and less than 0.5 % ash and no acid is a suitable agent for curing rubber. Furthermore, sulfur must be of medium fineness to disperse easily in the compound. For uniform vulcanization and optimum mechanical properties, sulfur must be dispersed uniformly throughout the rubber matrix. Sulfur dissolves in NR and SBR at room temperature, but not so easily in BR and NBR. The solubility of sulfur in rubber depends on its solubility coefficient. In NR, SBR, and BR, the solubility of sulfur increases as the temperature of the compound rises, producing an over-saturated solution. The amount of sulfur dissolved in the rubber depends on the mixing temperature. After mixing, the compound is kept in storage at an ambient temperature. In storage, sulfur crystallizes out of the over-saturated solution, forming crystals inside or on the surface of the compound. The latter is called blooming, and it depends on the over-saturation of the compound. Sulfur blooming is highly undesirable because it contaminates the surface of the compound and makes bonding problematic. To avoid blooming in unvulcanized rubber compounds and vulcanized rubber, the sulfur is added at the lowest possible temperature. To prevent sulfur blooming, insoluble sulfur with about 65-95% of material that is insoluble in rubber is used. At room temperature and even faster high temperatures, insoluble sulfur is unstable and slowly changes into the normal soluble form, and blooms again. The amount of sulfur required to cure rubber depends largely on the formulation of the compound. When the loading of sulfur is increased in a formulation, the degree of vulcanization or crosslinking rises too until the optimum is reached. The hardness continues to improve, but the overall mechanical properties, such as elongation at break, tensile strength, resilience, and ageing properties deteriorate. The number of sulfur atoms needed to create a crosslink depends on the type of accelerator and activating chemicals used. In typical soft rubber formulations, 0.2 to 5 parts per hundred rubber (phr) by weight of sulfur are used. When the loading of sulfur increases above 5 phr, the rubber becomes harder because of the excessive chemical crosslinks. Ebonite requires 25 to 40 phr of sulfur and there is no interest in using sulfur loading from 5 phr to 25 phr for most purposes [1.9].

1.3.2 Vulcanization with sulfur donors

Apart from sulfur, organic compounds that release sulfur at high temperatures can also be used to vulcanize rubber. Tetramethyl thiuram disulphide (TMTD) is one such compound. TMTD alone without sulfur does not liberate free sulfur during or after the vulcanization of compounds, and therefore its ability as a sulfur donor is open to dispute. But recent analytical methods have indicated that sulfur bridges are formed during crosslinking with TMTD. TMTD has also been studied in sulfurless vulcanization [1.9]. There are other thiuram disulphides or thiuram tetrasulphides available that are suitable for vulcanization with little or no sulfur. For instance, tetraethyl thiuram disulphide and dimethyl diphenyl thiuram disulphide must be used in larger quantities than the tetramethyl derivatives because of their higher molecular weight. They cure more slowly and are hence not generally used for sulfurless vulcanization [1.10]. In vulcanization, the sulfur-donor frees part of its loosely bound sulfur that forms crosslinks in the rubber. There are numerous advantages when sulfur donors are used. They reduce the normal blooming of sulfur in unvulcanized compounds, and the onset of cure occurs later than with pure sulfur, but when the reaction starts, the vulcanization proceeds quickly [1.9]. Vulcanization with TMTD is termed "thiuram vulcanization". Thiuram vulcanization gives excellent resistance to rubber at high temperatures and is used to make products with unsaturated rubber such as seals and cables. Normally, 2.5 phr to 3.5 phr TMTD is used without sulfur in rubber formulations. In many applications where TMTD is used, a small amount of sulfur, i.e., 0.1 phr to 0.3 phr, and an additional amount of a thiazole accelerator are added to improve the properties of the vulcanizates. The addition of a sulfenamide accelerator, as compared with pure thiuram vulcanization, helps to reduce scorch and bloom. Whether sulfur is used or not, ZnO is always incorporated with thiuram accelerators [1.10].

1.3.3 Chemical mechanism of thiuram disulphide crosslinking

It has always been assumed that thiuram disulphides are sulfur donors that form their corresponding monosulphides by splitting off one atom of sulfur. Although this was disputed in the past, in recent years, this view has been upheld once again. Tetramethyl thiuram disulphide disproportionate and tetramethyl thiuram monosulphide and trisulphide are produced. The latter splits off S_2, which then participates in the crosslinking as a bridge member. It is believed that dithioether bridge members are formed during thiuram vulcanization. Thiuram crosslinking never happens without zinc oxide participation in this reaction. Under the action of hydrogen sulphide, the hypothetic thiuram trisulphide is split up into dimethyl dithiocarbamic acid and radical sulfur, and this results in dithiocarbamic acid and zinc oxide forming zinc dithiocarbamate [1.10]. The reaction mechanism that assumes the presence of hydrogen sulphide has been rejected by some workers, who believe that the first and rate determining step is the splitting of the thiuram disulphide molecule into two dithiocarbamate acid radicals. They consider that the tetramethyl thiuram disulphide splits into the two radicals according to a first order time law. Of the dithiocarbamic acid formed, the amounts found again are 66% with the vulcanization of NR and 71% with the vulcanization of cis-polyisoprene. The zinc dithiocarbamate is also formed according to a time law of the first order, but more slowly. It is believed that an intermediate compound is formed from the rubber chain and the dithiocarbamate, and zinc dimethyl dithiocarbamate is formed during the vulcanization. Moreover, it has been shown that a reaction between tetramethyl thiuram disulphide and zinc oxide in a test tube results in 90% conversion to zinc dithiocarbamate, whereas the same reaction in rubber results only in 66% conversion. When vulcanization took place with thiuram disulphide and sulfur in the ratio of 1:6, 90% of the original amount of thiuram disulphide in the form of zinc dithiocarbamate was found. This indicated no loss of thiuram disulphide through the incorporation into the polymer chain [1.10].

1.4 Chemical curatives in sulfur vulcanization

1.4.1 Accelerators

Sulfur vulcanization of NR requires relatively long times at high temperatures with sulfur alone or with sulfur and inorganic accelerators. This process produces vulcanizates with poor resistance to ageing and shows severe sulfur blooming. There is also a high risk of over-curing. The mechanical properties deteriorate beyond the vulcanization optimum and the storage stability of the rubber articles and their resistance to ageing are poor, particularly at high temperatures [1.7,1.9]. Inorganic accelerators such as magnesium oxide, calcium hydroxide, lead oxide, or antimony tri- and pentasulfide shorten the vulcanization time and produce vulcanizates with better mechanical properties and longer service life when added with sulfur. However, they are no longer suitable for rubber, and the organic accelerators have replaced them. Organic accelerators perform several functions in the vulcanization of rubber. They shorten the curing times, increase the rate of crosslinking reaction with sulfur significantly and improve the ageing stability of the rubber products. A combination of different accelerators promotes a wider adjustment of rates and the onset of vulcanization. The sulfur content necessary to achieve optimum vulcanization properties can be lowered by adding accelerators. The organic accelerators are: dithiocarbamates; ammonium salts, metal salts; Xanthates, Thiurams; thiuram monosulphides, thiuram disulphide; Thiazoles; mercapto accelerators, sulfenamide accelerators; Aldehyde amine accelerators, Basic accelerators; guanidine accelerators, other basic accelerators [1.9]. The thiazoles are extremely important accelerators. The products of the thiazole accelerators, namely 2-mercaptobenzothiazole and its derivatives are the most popular accelerators. The thiazole accelerators are divided into mercapto accelerators and benzothiazole sulfenamide accelerators. Compounds with mercapto accelerators are suitable for all types of vulcanization at quite high temperatures with short cure times. None of the mercapto accelerators is fully effective unless ZnO is used. In the absence of a secondary accelerator, it is best to use fatty acids to improve the efficiency of the accelerator still further. This also helps to control the onset of vulcanization to a large extent and are used extensively because they produce excellent vulcanizate properties. To improve the efficiency of the thiazole accelerators, guanidines, thiurams, and dithiocarbamates are used with them [1.9]. The benzothiazole sulfenamide accelerators are derived from 2-mercapto-benzothiazole (MBT) because an amine is oxidatively bound to the mercapto sulfur. The accelerator depends on the type of amine used. Typical sulfenamide accelerators are N-cyclohexyl-2-benzothiazole sulfenamide (CBS), N-tert. butyl-2-benzothiazole sulfenamide (TBBS), 2-benzothiazole-N-sulfene morpholide (MBS), N,N-dicyclohexyl-2-benzothiazole sulfenamide (DCBS), and tert. Amyl-2-benzothiazole sulfenamide (5BS). In these accelerators, a molecular combination of mercapto accelerators and bases occurs, which makes the accelerators active as the amines are split off during vulcanization. The benzothiazole sulfenamides produce a retarded vulcanization start compared to the mercapto accelerators. The benzothiazole sulfenamides are activated with ZnO, and stearic acid is added to achieve higher crosslink densities in the compounds [1.11].

1.4.2 Activators

Organic accelerators can reach their full potential with organic or inorganic activators. ZnO is the most important inorganic additive in this respect. The cure system is further activated by adding organic activators which are fatty acids, for example, stearic, palmitic, and lauric acid, weak amines, the salts of these two groups of chemicals, and polyalcohols and amino alcohol, such as ethylene glycol and triethanolamine [1.12].

A similar activity is also observed with dibutylaminooleate, 1, 3-diphenylguanidine-phthalate and amines, similar to mono-, di- and triethanolamine, mono-, and dibutylamine, and dibenzylamine. The basic activators help to shorten the vulcanization time. In addition, the fatty acids and fatty acid salts give better processing and more homogenous dispersion of solid fillers and chemicals in the rubber [1.11].

1.4.3 Retarders

When vulcanization times are short, or processing temperatures are high, one often must retard the onset or start of vulcanization to assure enough processing safety. There are chemicals that retard vulcanization, such as N-nitroso compounds of secondary aromatic amines, for example, N-nitrosodiphenylamine (NDPA). For safety reasons, these compounds are no longer used. Other retarders are low volatility organic acids, such as benzoic acid (BES), phthalic anhydride (PTA), and salicylic acid (SCS). Some more recent retarders are sulfenic acid and sulfonic derivatives that have gained importance [1.11].

1.5 Various sulfur vulcanization systems

As mentioned before, sulfur is the most important chemical for the vulcanizing of soft rubber goods, and dosages of 0.25-5.0 phr are used in most formulations. For ebonite, the dosage can be as high as 25 phr to 40 phr. The amount of sulfur used for making soft rubber articles depends on the type and loading of accelerators used and the demand for properties of the vulcanizate. For less active accelerators, larger amounts of sulfur are needed, i.e., bases like guanidine, and for highly active accelerators, for example, sulfenamides, less sulfur is required. The most widely used system is the conventional, where 1.5-2.5 phr sulfur with 1.0-0.5 phr accelerator is employed. The sulfur to accelerator weight ratio is greater than 1, and it leads to mostly polysulfidic crosslinks in the rubber. The same crosslink density can be achieved by lowering the sulfur content and increasing the accelerator dosage. When sulfur content is reduced to about 1.2-0.5 phr and the accelerator dosage is raised to 1.5-2.5 phr, this produces crosslinks with less sulfur content and is termed a semi-efficient (semi-EV) system. In this system, the sulfur to accelerator weight ratio is less than 1. Rubber products cured with semi-EV are heat- and reversion-resistant vulcanizates. The high accelerator loading is a more effective use of sulfur and produces shorter sulfur links. When the quantity of accelerators is increased significantly, small amounts of sulfur are needed to achieve a good degree of crosslinking. At very low dosages of sulfur, accelerators that are sulfur-donors (e.g. TMTD) are used. For instance, 2.5-3.5 phr TMTD and a very small amount of sulfur, as low as 0.2 phr, are used for vulcanization. In some cases, only the sulfur-donor accelerators are used in the absence of sulfur for vulcanization. The efficient vulcanization (EV) system is when the sulfur to accelerator weight ratio reaches 1. In the EV system, the crosslink structure consists of mono- and di-sulfidic crosslinks. Rubbers cured with the EV system, have very good heat- and reversion-stability, and a low high-temperature compression set [1.11]. The cure systems in industrial rubber articles vary depending on their applications. It is common to tailor the cure system to achieve a certain effect on the properties of the article. For example, in radial truck tread formulation, the sulfur to accelerator ratio is 1, making it an EV system. In the conventional passenger tread formulation, the sulfur to accelerator ratio is 2, making it a conventional system, whereas in the passenger tread of the green tire formulation, the sulfur to accelerator ratio is less than 1, making it a semi-EV system [1.13]. Other rubber articles have also benefited from conventional and semi-EV systems. For example, viper blades have a conventional system, whereas, conveyor belt covers, footwear, and general gaskets have a semi-EV system for vulcanization [1.14].

There is an impressive range of chemicals that cause sulfur to react with rubber efficiently to produce crosslinks. The sulfur vulcanization of rubber is here to stay, but the harmful effects of chemical curatives and the danger they pose can no longer be ignored. It is time to address these concerns and make sulfur vulcanization safer, cheaper, and environmentally friendly. It is time to take a step towards green rubber products.

1.6 References

[1.1] – L. Bateman, "The chemistry and physics of rubber-like substances", Studies of the natural rubber producers' research association, Maclaren & Sons Ltd. London, John Wiley & Sons. New York. 1963, chapter 1.

[1.2] – C. Hepburn, R. J. W. Reynolds, "Elastomers: Criteria for engineering design", Applied Science Publishers Ltd, London, 1979, Chapters 1&12. (ISBN 0-85334-809-X)

[1.3] – https://www.deltarubber.co.uk/rubber-industry Date visited 30-11-2021

[1.4] – S. C. Nyburg, "X-ray determination of crystallinity in deformed natural rubber ", Brit J. Appl. Phys., 1954, 5, 321.

[1.5] – L. Bateman, "The chemistry and physics of rubber-like substances", Studies of the natural rubber producers' research association, Maclaren & Sons Ltd. London, John Wiley & Sons. New York. 1963, chapter 9.

[1.6] – P. A. Ciullo, N. Hewitt, "The rubber formulary", Noyes Publications, NY 1999.

[1.7] – W. Hofmann, "Rubber technology handbook," Hanser Publishers, New York, 1989, Chapter 3. (ISBN 3-446-14895-7) (Hanser) Pp.

[1.8] – A. D. Roberts, "Natural rubber science and technology", Oxford science publications, 1988, Chapter 19 (ISBN 0-19-855225-4).

[1.9] – W. Hofmann, "Vulcanization and vulcanizing agents", Maclaren and Sons Ltd, London, 1967, Chapter 2.

[1.10] – W. Hofmann, "Vulcanization and vulcanizing agents", Maclaren and Sons Ltd, London, 1967, Chapter 3.

[1.11] – W. Hofmann, "Rubber technology handbook," Hanser Publishers, New York, 1989, Chapter 4. (ISBN 3-446-14895-7) (Hanser) Pp.

[1.12] – W. Hofmann, "Vulcanization and vulcanizing agents", Maclaren and Sons Ltd, London, 1967, Chapter 4.

[1.13] – R. N. Datta, "Rubber curing systems", Rapra Review Report 144, Rapra Technology Ltd, 12 (12), 2002, pp 32-37. (ISSN: 0889-3144)

[1.14] – P. A. Ciullo, N. Hewitt, "The rubber formulary", Noyes publications, New York, 1999, pp 85, 91,147, and 275.

CHAPTER 2

2 High efficiency sulfur vulcanization of rubber with fewer chemicals

2.1 Method 1 – Use of one accelerator, one activator, and sulfur to cure NR, BR and EPDM rubbers

2.1.1 Introduction

Since Charles Goodyear discovered in 1843 that heating raw rubber with sulfur modified the rubber to retain its shape, there has been an increasing trend to use chemical additives to improve the processing, mechanical, and dynamic properties of raw rubber. The use of two accelerators and two activators in combination with sulfur in cure systems in industrial rubber compounds is widespread. For instance, in carbon black-filled NR-based tire belt skim compound, the cure system consists of 5 phr sulfur, two activators: 7 phr ZnO and 1 phr stearic acid, and two accelerators: 2 phr hexa-methoxymethyl-melamine (HMMM) and 0.7 phr N,N-Dicyclohexyl-2-benzothiazole-sulfenamide (DCBS), adding up to 15.7 phr [2.1]. Excessive use of chemical curatives is harmful to health, safety, and the environment, and their use is restricted by the new European chemicals policy, Registration, Evaluation, Authorization, and Restriction of Chemicals (REACH) and various laws for the environment and safety. For example, zinc oxide (ZnO) is used as a primary activator in sulfur-cure systems with accelerators, sulfur, and fatty acids, e.g., stearic acid, to form a complex that serves as the actual accelerating agent. The presence of ZnO and stearic acid in sulfur-cure systems has come under increasing scrutiny because of environmental concerns. The European Union in the Commission Directive 2004/73/EC classified ZnO as "dangerous for the environment" and "very toxic to aquatic organisms." Zinc oxide may also cause long-term adverse effects on the aquatic environment. Stearic acid is highly flammable and irritating to the eyes, respiratory system, and skin [2.2]. The chemicals in the cure systems have not been measured accurately, and there is no reason why so many of them are used in the curing of rubber. In addition, when too many chemicals are mixed with raw rubber, they may not disperse and distribute well in the rubber matrix. When the rubber compound is cured at a high temperature, some chemicals melt and re-aggregate within the rubber when stored at ambient temperature. The aggregates may eventually migrate through the rubber to the surface and cause contamination. In a study, the effect of the re-agglomeration and migration of some chemical curatives on the mechanical properties of an NR vulcanizate reinforced with a silane pre-treated precipitated silica was examined. The cure system was made of 6 phr TBBS and 0.3 phr ZnO. The sulfur in the organosilane was used to vulcanize the rubber. The rubber compound was cured at temperatures below and above the melting point of the TBBS (105°C) and stored at ambient temperature (21±2°C) for up to 60 days to allow full blooming to appear on the rubber surface. The mechanical properties of the rubber vulcanizate were measured and found to be affected by the blooming of the TBBS on the rubber surface, which also damaged the internal structure of the rubber. When the rubber surfaces were analysed in a Fourier transform infrared spectrometer (FT-IR-8400S), it appeared that below the melting point of the TBBS, there was little or no re-agglomeration and some limited migration of the TBBS, whereas above the melting point, there was re-agglomeration and extensive migration of the TBBS to the rubber surface. The cyclic fatigue life increased when the extent of the migration of the TBBS to the rubber surface was reduced. There was some improvement in the fatigue properties when the re-agglomeration of the TBBS in the rubber was absent [2.3]. The evidence suggested that during high temperature curing, some chemicals remained unreacted and migrated to the surface when the rubber was kept in storage at room temperature. This caused the surface contamination. A reduction in the use of chemical curatives will reduce the blooming and make the rubber more durable in service.

Two highly efficient methods have been developed to measure the exact amounts of ZnO, and sulfenamide and thiuram accelerator requirements for the sulfur vulcanization of NR, BR, and EPDM rubbers. One method uses one accelerator, one activator, and sulfur, and the other method, uses two single additives to cure rubber with and without additional sulfur. Using fewer chemicals in the cure system did not compromise the cure efficiency that is important in an industrial manufacturing setting. Both methods are more efficient than the current systems for sulfur vulcanization.

2.1.2 Materials and Mixing

The raw rubbers used were standard Malaysian natural rubber (NR) grade L (98 wt % 1, 4-cis content; SMRL), high cis polybutadiene rubber (96 wt% 1,4-cis content, Buna CB24, Bayer, Newbury, UK, not oil extended), and ethylene-propylene-diene rubber (EPDM, 48 wt% ethylene content, 9 wt% ethylidene norbornene content, and 13 wt% oil content, Keltan 6251A, Lanxess, The Netherlands). The other ingredients were sulfur (curing agent: Solvay Barium Strontium, Hanover,

Germany), N-tert-butyl-2-benzothiazole sulfenamide (a fast-curing delayed action accelerator with a melting point of 105°C) (Santocure TBBS, Sovereign Chemicals, USA), tetramethyl thiuram disulphide (a fast curing sulfur-donor accelerator with approximately 13 wt% of the sulfur available to react with rubber and a melting point of 156°C) (TMTD, Sigma-Aldrich, UK), zinc oxide (ZnO; an activator, Harcros Durham Chemicals, Durham, UK), and stearic acid (an activator, Anchor Chemicals Ltd., UK). These chemicals are used extensively in industrial rubber compounds [2.4]. The chemical structures of TBBS and TMTD accelerators are shown in Scheme 2.1.

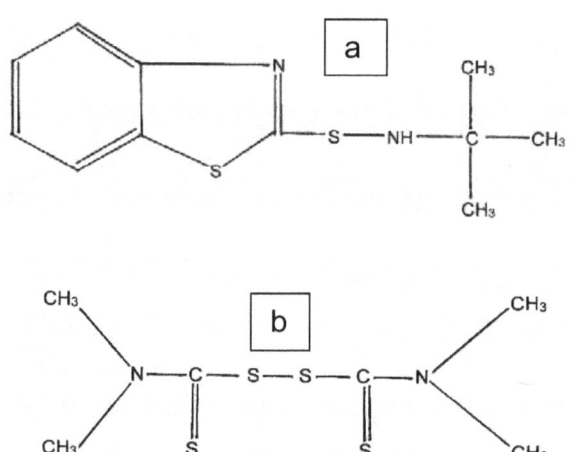

Scheme 2.1 – Chemical structures of the accelerators. a) TBBS, b) TMTD

The raw rubber was mixed with the chemical ingredients in a Haake Rheocord 90 (Berlin, Germany), a small size laboratory mixer with counter-rotating rotors to produce compounds. The Banbury rotors and the mixing chamber were initially set at room temperature (23°C), and the rotor speed was set to 45 r.p.m. The volume of the mixing chamber was 78 cm³, and it was 60% full during mixing. Polylab monitor 4.17 software program was used for controlling the mixing conditions and storing data. To prepare the rubber compounds, the raw rubber was introduced first into the mixer, and after 30 seconds, the sulfur, TBBS, ZnO, and stearic acid were added and mixed for 8 min. This procedure was repeated for making all the rubber compounds. Over three hundred rubber compounds were made and tested. The temperature of the rubber compounds during mixing reached 48-62°C. Each point in the figures presented here corresponds to one rubber compound.

2.1.3 Testing of the rubber compounds

The viscosity of the rubber compounds was measured at 100°C in a single-speed rotational Mooney viscometer (Wallace Instruments, Surrey, UK) according to British Standard 1673, Part 3, 1969, and the results were expressed as Mooney Units (MU). The cure properties of the rubber compounds were determined using an oscillating disc rheometer curemeter (ODR, Monsanto, Swindon, UK) at 160±2°C with an angular displacement of ± 3° and a test frequency of 1.7 Hz in accordance with British Standard ISO 3417, 2008. The tests ran for up to 2 hours. The cure properties measured were: scorch time, t_{s2}; optimum cure time, t_{95}; cure rate index, CRI; maximum torque, M_H; and minimum torque, M_L. ΔTorque ($M_H - M_L$) is the difference between the maximum and minimum torques on the cure trace of a rubber compound and is an indication of crosslink density changes. Δtorque was then plotted against the loading of TBBS, ZnO, and stearic acid to measure the optimum amount of the additives required for curing the rubber. The viscosities of the raw NR, BR, and EPDM rubbers were 89, 46, and 88 MU, respectively.

2.1.4 Measurement of the optimum amounts of TBBS and ZnO for curing NR

The rubber compounds had 1 phr, 2 phr, 3 phr, and 4 phr sulfur. Firstly, TBBS was added. The loading of TBBS was increased progressively from 0.25 phr to 4.5 phr to measure the optimum amount required to cause a reaction between sulfur and the rubber (35 rubber compounds were made). Then, ZnO was added to improve the effectiveness of the TBBS during curing. The loading of ZnO in the rubber with sulfur and TBBS was raised from 0.05 phr to 0.5 phr (32 rubber compounds were made). The effects of stearic acid on the cure properties of the rubber compounds with sulfur, TBBS, and ZnO were measured (8 more rubber compounds were prepared). The loading of stearic acid was raised from 0 to 2.5 phr.

Figure 2.1 shows Δtorque as a function of TBBS loading for the rubbers with 1 phr, 2 phr, 3 phr, and 4 phr sulfur. For the rubber with 1 phr sulfur, Δtorque increased from 5 dNm to 23 dNm as the loading of TBBS was raised to 1.5 phr. It then continued rising at a much slower rate, reaching about 32 dNm when the loading of TBBS reached 3.5 phr. The addition of 1.5 phr TBBS was enough to cause the sulfur to react with the rubber to form covalent sulfur crosslinks. Similarly, for the rubbers with 2 phr and 3 phr sulfur, 1.5 phr TBBS was enough to react the sulfur with the rubber. Further increases in the TBBS loading to 3.5 phr had a lesser effect on the Δtorque, which increased only marginally. For the rubber with 4 phr sulfur, Δtorque increased from 24 dNm to 41 dNm as the loading of TBBS was increased from 0.5 phr to 3.5 phr. It then rose to 43 dNm when an additional 1 phr TBBS was incorporated into the rubber. The 3.5 phr TBBS was enough to react the sulfur with the rubber. The TBBS requirement in the cure system remained essentially constant at 1.5 phr when the loading of sulfur was increased to 3 phr, but it rose to 3.5 phr with 4 phr sulfur.

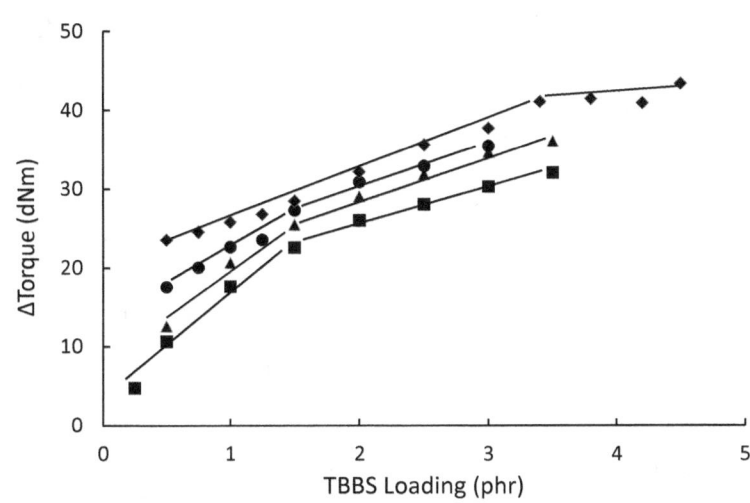

Figure 2.1 - ΔTorque vs. TBBS loading for the rubber compounds with different amounts of sulfur. (■) with 1 phr sulfur; (▲) with 2 phr sulfur; (●) with 3 phr sulfur; (♦) with 4 phr sulfur.

After these measurements were completed, four rubber compounds were selected (Table 2.1).

Table 2.1: Rubber formulations and cure properties of the NR rubber compounds with the optimum amounts of TBBS.

Formulation (phr)	Compound no.			
	1	2	3	4
SMRL (NR)	100	100	100	100
Sulfur	1	2	3	4
TBBS	1.5	1.5	1.5	3.5
ODR test results at 160°C				
M_L (dNm)	14	15	15	14
M_H (dNm)	37	40	42	55
ΔTorque (dNm)	23	25	27	41
t_{s2} (min)	10	8.5	7.5	7.5
t_{95} (min)	16	14	12.2	13.5
CRI (min^{-1})	16.7	18.2	21.3	16.7

The cure traces of the rubber compounds in Table 2.1 are presented in Fig. 2.2. It is interesting that the rubber compounds were cured with only TBBS and no ZnO and no stearic acid.

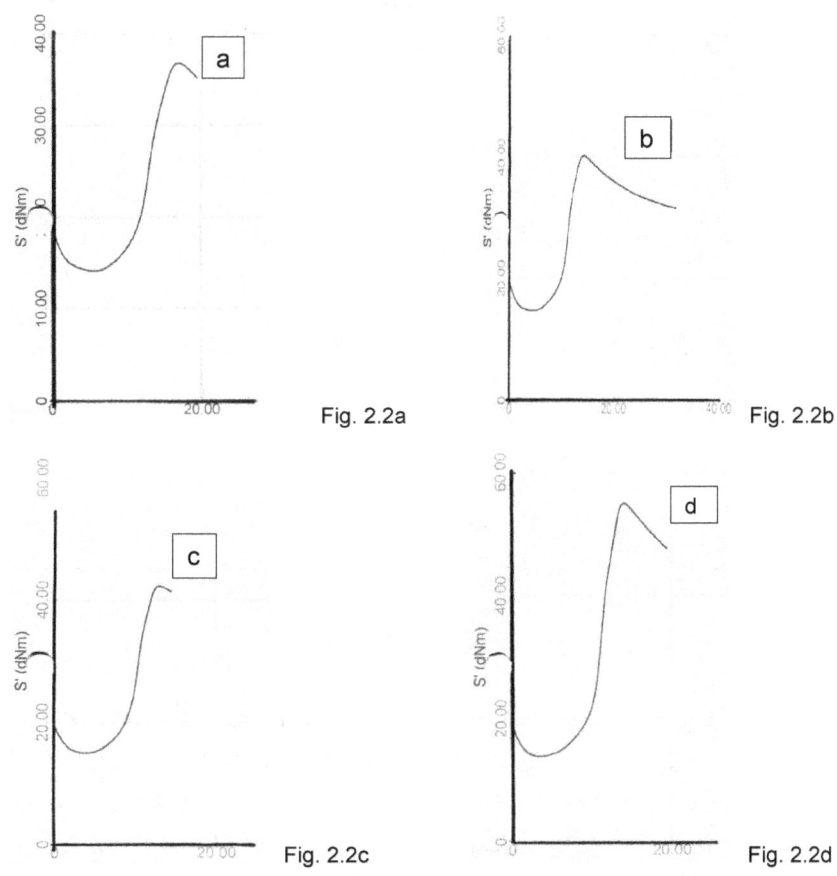

Fig. 2.2a

Fig. 2.2b

Fig. 2.2c

Fig. 2.2d

Figure 2.2 - Cure traces [(torque vs. time (min)] of the rubber compounds with different amounts of sulfur and TBBS. a) with 1 phr sulfur and 1.5 phr TBBS; b) with 2 phr sulfur and 1.5 phr TBBS; c) with 3 phr sulfur 1.5 phr TBBS; d) with 4 phr sulfur and 3.5 phr TBBS.

To enhance the efficiency of the reaction of the TBBS with sulfur, ZnO was added. The addition of 0.2-0.3 phr ZnO to the rubbers with 1 phr, 2 phr, and 3 phr sulfur and 1.5 phr TBBS, improved the efficiency of the TBBS, as shown by a large increase in the Δtorque values. Interestingly, the rubber compound with 4 phr sulfur and 3.5 phr TBBS needed 0.2 phr ZnO to improve the crosslink density, as indicated by an increase in Δtorque. The ZnO requirement was not affected by higher loadings of TBBS and remained very low at 0.2-0.3 phr (Fig. 2.3).

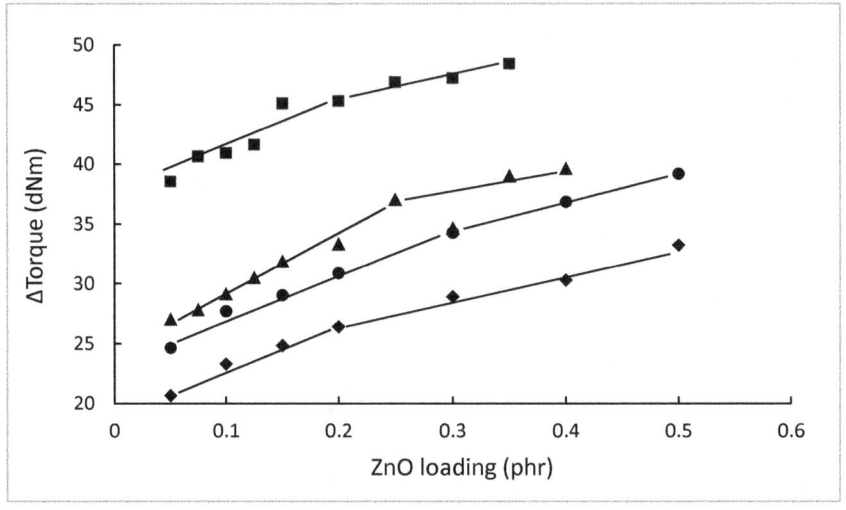

Figure 2.3 - ΔTorque vs. ZnO loading for the rubber compounds with different amounts of sulfur and TBBS. (♦) with 1 phr sulfur and 1.5 phr TBBS; (●) with 2 phr sulfur and 1.5 phr TBBS; (▲) with 3 phr sulfur and 1.5 phr TBBS; (■) with 4 phr sulfur and 3.5 phr TBBS.

After these measurements were completed, four rubber compounds were selected (Table 2.2).

Table 2.2: Rubber formulations and cure properties of the NR rubber compounds with the optimum amounts of TBBS and ZnO.

Formulation (phr)	Compound no.			
	1	**2**	**3**	**4**
SMRL (NR	100	100	100	100
Sulfur	1	2	3	4
TBBS	1.5	1.5	1.5	3.5
ZnO	0.2	0.3	0.25	0.2
ODR test results at 160°C				
M_L (dNm)	16	17	16	14
M_H (dNm)	42	52	53	58
ΔTorque (dNm)	26	35	37	44
t_{s2} (min)	8.4	6.5	5.6	4.9
t_{95} (min)	15.5	10.7	9.8	9.4
CRI (min^{-1})	14.1	23.8	23.8	22.2
Exact cure system (S/TBBS/ZnO)	(1/1.5/0.2)	(2/1.5/0.3)	(3/1.5/0.25)	(4/3.5/0.2)

The cure traces of the rubber compounds in Table 2.2 are presented in Fig. 2.4. It is interesting that the rubber compounds were cured with TBBS and ZnO and no stearic acid.

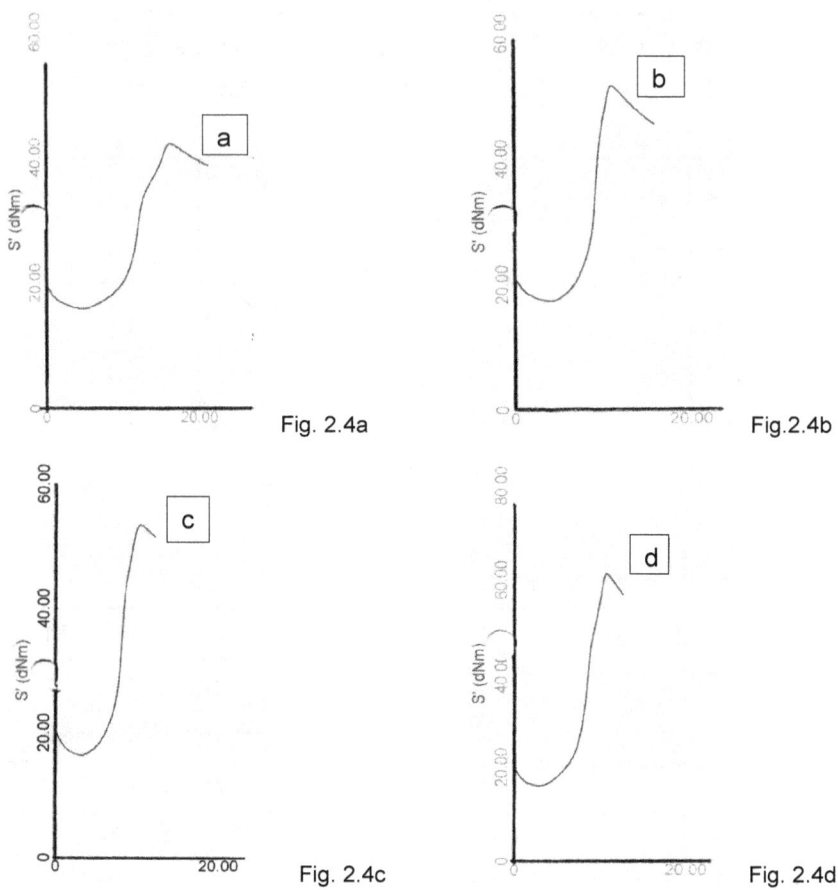

Fig. 2.4a

Fig.2.4b

Fig. 2.4c

Fig. 2.4d

Figure 2.4 – Cure traces [(torque vs. time (min)] of the rubber compounds with different amounts of sulfur, TBBS and ZnO. a) with 1 phr sulfur, 1.5 phr TBBS and 0.2 phr ZnO; b) with 2 phr sulfur, 1.5 phr TBBS and 0.3 phr ZnO; c) with 3 phr sulfur, 1.5 phr TBBS and 0.2 phr ZnO; d) with 4 phr sulfur, 3.5 phr TBBS and 0.2 phr ZnO.

The fact that a small amount of ZnO, i.e., as low as 0.2 phr, improved the performance of the TBBS during curing in the absence of stearic acid is interesting. The sulfur-cure system in tire belt skim compound, contains up to 7 phr ZnO, which is far too excessive [2.1]. As mentioned before, stearic acid is a fatty acid and is added as a co-activator with ZnO in sulfur vulcanization to enhance the efficiency of the curing reaction in the rubber. Stearic acid acts as a plasticizer and an internal lubricant between polymer chains and aids dispersion and solubility of solid ingredients like ZnO in the rubber. It is assumed that stearic acid reacts with ZnO to form zinc stearate, which is an essential cure activator [2.5]. The rubber compound with 1 phr sulfur, 1.5 phr TBBS, and 0.2 phr ZnO was mixed with stearic acid. The loading of stearic acid was raised from 0 to 2.5 phr (Fig. 2.5). The inclusion of stearic acid in the cure system had no benefit for the Δtorque which stayed constant at about 27 dNm.

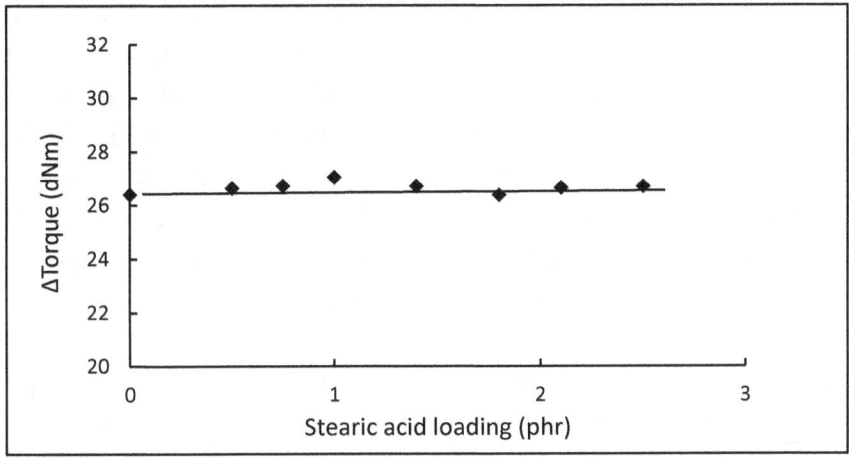

Figure 2.5 – ΔTorque vs. stearic acid loading for the rubber compound with 1 phr sulfur, 1.5 phr TBBS and 0.2 phr ZnO.

This method measured the exact amounts of the TBBS and ZnO for an optimum cure at different loadings of sulfur. For example, at 1 phr sulfur loading, the TBBS and ZnO requirements for optimum cure were 1.5 phr and 0.2 phr, respectively. At 4 phr sulfur loading, the TBBS requirement changed to 3.5 phr, but that of ZnO remained at 0.2 phr. The

TBBS and ZnO requirements were not affected by the loading of sulfur up to 3 phr.

An increase in the loading of sulfur from 1 phr to 4 phr had a big influence on the scorch and optimum cure times and the cure rate. The scorch and optimum cure times shortened from 8.4 to 4.9 min, and 15.5 to 9.4 min, respectively, and the crosslink density, as indicated by the rise in the Δtorque value, increased from 26 dNm to 44 dNm. The CRI, which indicates the cure rate, rose from 14.1 min^{-1} to 23.8 min^{-1} when up to 3 phr sulfur was added, and decreased marginally to about 22 min^{-1} when the loading of sulfur reached 4 phr (Table 2.2). The rubber compounds were cured fully, as shown in Fig. 2.4, despite using fewer chemicals. Contrary to the common practice, a high level of cure efficiency can be achieved with fewer chemicals. For example, there is 97 wt% less ZnO, no stearic acid, 44 wt% less accelerator, and up to 80 wt% less sulfur in compound 1, and 20 wt% less sulfur in compound 4 than there is in the NR-based tire belt skim compound formulation [2.1]. This is a significant cost-saving and improves health and safety at work and causes less damage to the environment.

The method used traces from a curemeter to measure the Δtorque vs TBBS and Δtorque vs ZnO correlations. The correlations were used to determine the optimum amounts of the TBBS and ZnO required to fully cure the NR with different amounts of sulfur. The following steps must be taken.

- Mix sulfur with the raw rubber and add TBBS to cause the sulfur to react with the rubber. Use the ΔTorque vs TBBS correlation to measure the optimum amount of TBBS required for curing the rubber.

- Add ZnO to the rubber with sulfur and the optimum amount of TBBS to improve the reaction of TBBS with sulfur. Use the ΔTorque vs ZnO correlation to determine the exact amount of ZnO required for optimum cure.

2.1.5 Measurement of the optimum amounts of TBBS and ZnO for curing BR

The rubber compounds had 0.5 phr and 1 phr sulfur. TBBS was added first. The loading of TBBS was increased progressively from 0.5 phr to 3.8 phr to measure the optimum amount required to cause a reaction between sulfur and the rubber (18 rubber compounds were made). Then, ZnO was added to improve the effectiveness of the TBBS during curing. The loading of ZnO in the rubber compounds with sulfur and TBBS was raised from 0.05 phr to 0.45 phr (18 rubber compounds were made). Seven more rubber compounds were prepared to evaluate the effect of stearic acid on the cure properties of the rubber compounds with sulfur, TBBS, and ZnO. Stearic acid was used as a secondary activator with ZnO. The loading of stearic acid was raised from 0 to 2.5 phr to measure its effects on the cure properties of the rubber compounds. After the rubber compounds were mixed, they were tested in a curemeter to produce traces from which the cure properties were determined. ΔTorque was plotted against the loading of TBBS, ZnO, and stearic acid.

Figure 2.6 shows Δtorque as a function of TBBS loading for the BR rubber compounds with 0.5 phr and 1 phr sulfur. For the rubber compound with 0.5 phr sulfur, Δtorque increased from 25 dNm to 45 dNm as the loading of TBBS was raised to 1.75 phr. It then stopped rising when the loading of TBBS reached 3 phr. The addition of 1.75 phr TBBS was enough to react the sulfur with the rubber to form crosslinks. For the rubber with 1 phr sulfur, Δtorque increased from 39 dNm to 61 dNm as the loading of TBBS was raised from 0.5 phr to 3 phr. It then remained essentially unchanged when an additional 0.8 phr

TBBS was incorporated into the rubber. The 3 phr TBBS was enough to react the sulfur with the rubber.

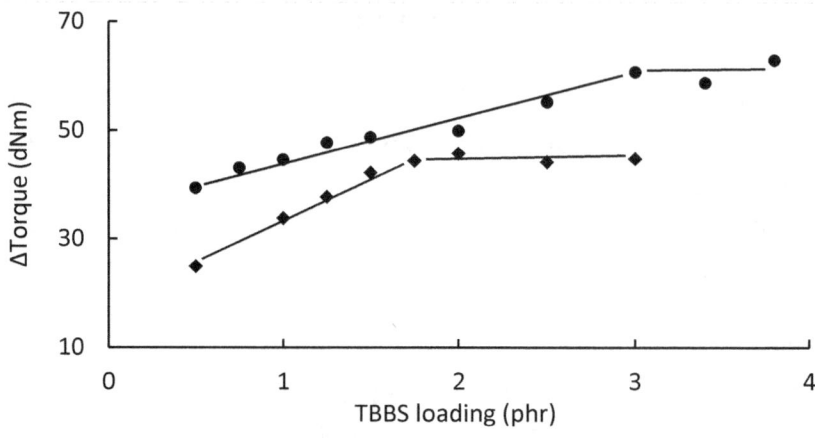

Figure 2.6 – ΔTorque vs. TBBS for BR with different sulfur loading. (◆) 0.5 phr sulfur, (●) 1 phr sulfur

ZnO was added to enhance the efficiency of cure in the BR with sulfur and TBBS (Fig. 2.7). For the rubber compound with 0.5 phr sulfur and 1.75 phr TBBS, the addition of 0.2 phr ZnO was enough to make the TBBS more effective, as indicated by an increase in Δtorque from 62 dNm to 71 dNm. When the loading of ZnO was raised to 0.35 phr, Δtorque showed no increase and remained at about 72 dNm. ΔTorque for the rubber with 1 phr sulfur and 3 phr TBBS reached a maximum value of 92 dNm at 0.2 phr ZnO. It then increased marginally to about 94 dNm when an extra 0.25 phr ZnO was added.

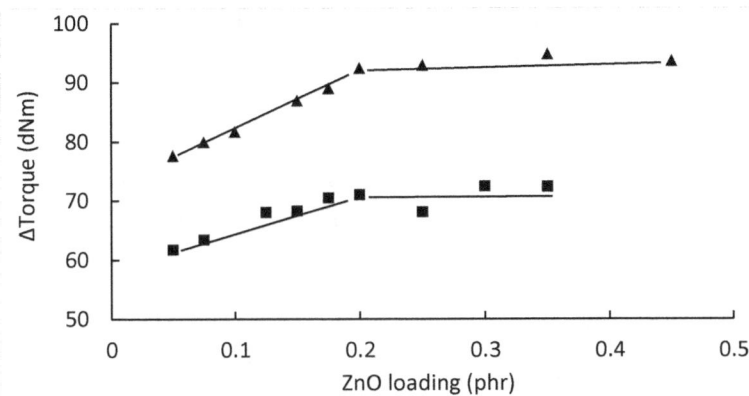

Figure 2.7 – ΔTorque vs. ZnO loading for BR with TBBS
and different amounts of sulfur. (■) 0.5 phr sulfur and
1.75 phr TBBS, (▲) 1 phr sulfur and 3 phr TBBS.

When 0.5 phr stearic acid was added to the BR with 0.5 phr
sulfur, 1.75 phr TBBS and 0.2 phr ZnO, Δtorque first decreased
from 71 dNm to 50 dNm and then remained steady as the
loading of stearic acid reached 2.5 phr. A small amount of
stearic acid, i.e., 0.5 phr, was detrimental to the crosslink
density, as indicated by a drop in Δtorque (Fig. 2.8).

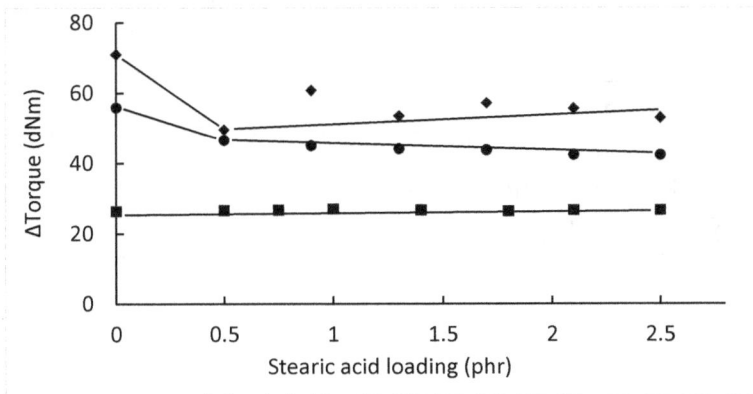

Figure 2.8 – ΔTorque vs. stearic acid loading for BR with
0.5 phr sulfur, 1.75 phr TBBS and 0.2 phr ZnO (♦), NR with
1 phr sulfur, 1.5 phr TBBS and 0.2 phr ZnO (■), EPDM
with 1 phr sulfur, 1 phr TBBS and 0.075 phr ZnO (●).

After these measurements were completed, two rubber compounds were selected (compounds 1 and 2, Table 2.3).

Table 2.3: Rubber formulations and cure properties of the BR and EPDM rubber compounds with the optimum amounts of TBBS and ZnO.

Formulation (phr)	Compound no.		
	1	2	3
EPDM	-	-	100
BR	100	100	-
Sulfur	0.5	1	1
TBBS	1.75	3	1
ZnO	0.2	0.2	0.075
ODR test results at 160°C			
M_L (dNm)	15	15	16
M_H (dNm)	85	108	72
ΔTorque (dNm)	70	93	56
t_{s2} (min)	13.2	11.8	17
t_{95} (min)	55.7	43.3	37.7
CRI (min^{-1})	2.4	3.2	4.8

The cure traces of the rubber compounds listed in Table 2.3 are shown in Fig 2.9.

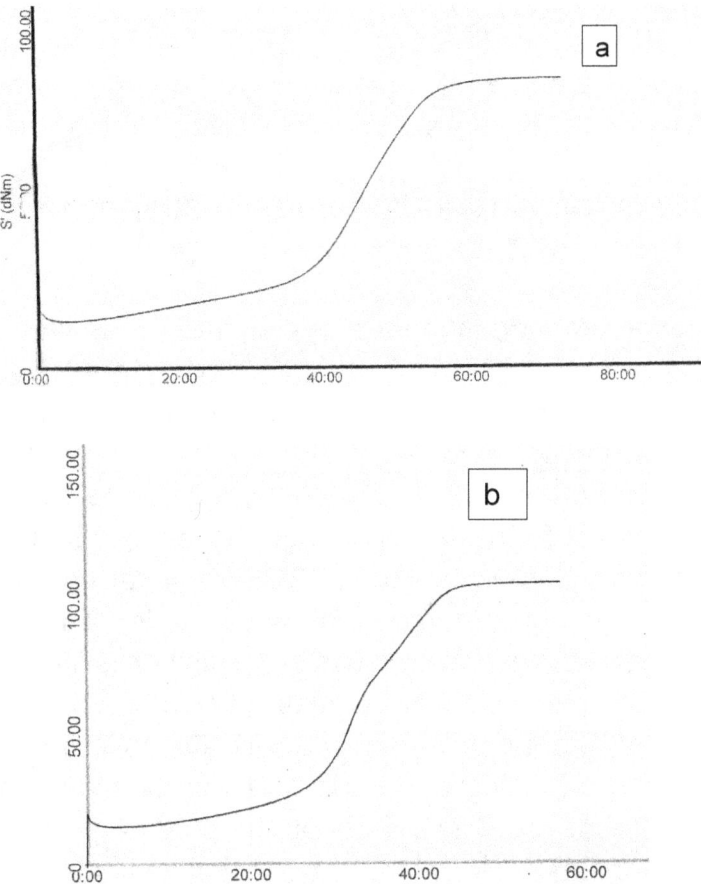

Figure 2.9 – Cure traces [(torque vs. time (min)] of the BR rubber compounds with different amounts of sulfur, TBBS and ZnO. a) with 0.5 phr sulfur, 1.75 phr TBBS, and 0.2 phr ZnO; b) with 1 phr sulfur, 3 phr TBBS and 0.2 phr ZnO.

This method measured the exact amounts of the TBBS and ZnO for an optimum cure at different loadings of sulfur. For example, at 0.5 phr sulfur loading, the TBBS and ZnO requirements for an optimum cure were 1.75 phr and 0.2 phr, respectively. At 1 phr sulfur loading, the TBBS requirement changed to 3 phr TBBS and that of ZnO remained at 0.2 phr. The ZnO requirement was not affected by changes in the amounts of sulfur and TBBS. An increase in the loading of sulfur from 0.5 phr to 1 phr improved the Δtorque by 27% and reduced the t_{s2} and t_{95} by 11% and 22%, respectively. The CRI increased by 33%. A higher loading of sulfur was beneficial to the cure cycle.

2.1.6 Measurement of the optimum amounts of TBBS and ZnO for curing EPDM

The rubber compound had 1 phr sulfur. TBBS was added first. The loading of TBBS was increased progressively from 0.25 phr to 3.8 phr to measure the optimum amount required to cause a reaction between sulfur and the rubber (9 rubber compounds were made). ZnO was then added to improve the effectiveness of TBBS during curing. The loading of ZnO in the rubber compounds with sulfur and TBBS was raised from 0 phr to 0.4 phr (11 rubber compounds were made). The effect of stearic acid on the cure properties of the rubber compound with sulfur, TBBS, and ZnO were measured (7 rubber compounds were made). Stearic acid was used as a secondary activator with ZnO. The loading of stearic acid was raised from 0 phr to 2.5 phr to measure its effects on the cure properties of the rubber compounds.

Figure 2.10 shows Δtorque vs. TBBS loading for the EPDM with 1 phr sulfur. ΔTorque increased from 26 dNm to 42 dNm as the loading of TBBS was raised from 0.25 phr to 1 phr. There was no improvement in Δtorque when the amount of TBBS reached 3.8 phr. The 1 phr TBBS was enough to cause the sulfur to react with the rubber.

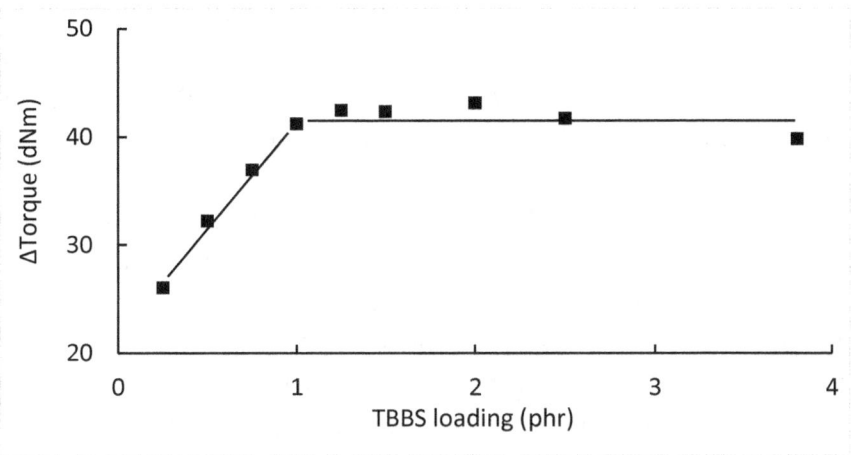

Figure 2.10 – ΔTorque vs. TBBS loading
for EPDM with 1 phr sulfur.

ZnO was added to improve the efficiency of the TBBS. ΔTorque rose from 41 dNm at 0 phr ZnO to 56 dNm at 0.075 phr ZnO. The rate of increase slowed significantly and Δtorque reached about 67 dNm when the loading of ZnO was raised by an additional 0.325 phr (Fig. 2.11).

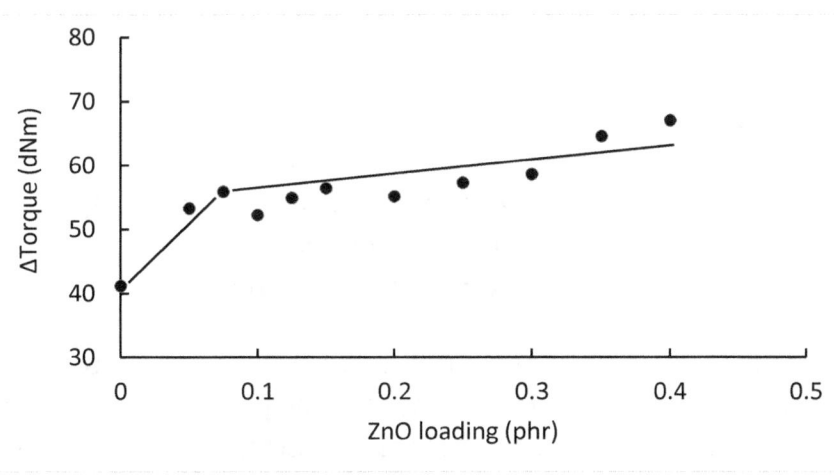

Figure 2.11 – ΔTorque vs. ZnO loading for
EPDM with 1 phr sulfur and 1 phr TBBS.

It is remarkable that a small amount of ZnO, i.e., as low as 0.075 phr, had such a big effect on the performance of TBBS in the cure system, as indicated by a significant rise in Δtorque. When up to 2.5 phr stearic acid was mixed with the EPDM with 1 phr sulfur, 1 phr TBBS, and 0.075 phr ZnO, Δtorque decreased from 56 dNm to 47 dNm with 0.5 phr stearic acid. It then continued decreasing at a slower rate to about 42 dNm when a further 2 phr stearic acid was added. The crosslink density was adversely affected by the stearic acid (Fig. 2.8). After these measurements were completed, one rubber compound was selected (Compound 3, Table 2.3). The trace from which the cure properties were measured is shown in Fig 2.12.

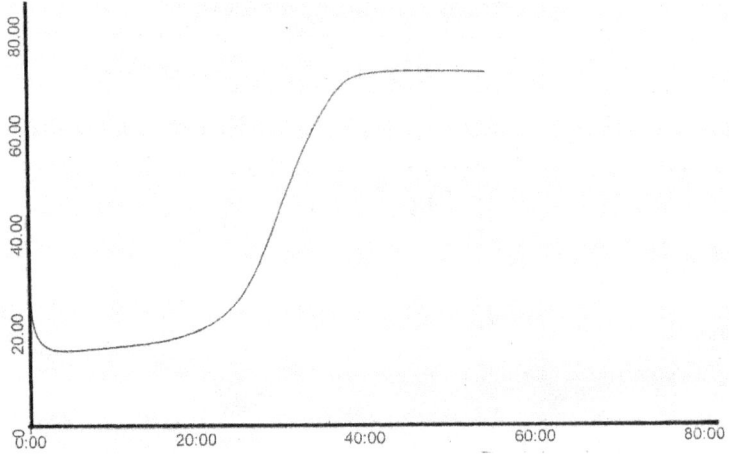

Figure 2.12 – Cure trace [(torque vs. time (min)] of the EPDM rubber compound with 1 phr sulfur, 1 phr TBBS, and 0.075 phr ZnO.

This method uses traces from a curemeter to measure the exact amounts of the TBBS and ZnO required for fully curing the NR, BR, and EPDM rubbers at different levels of sulfur very accurately. For the NR, the TBBS and ZnO requirements remained at 1.5 phr and 0.2-0.3 phr, respectively, when up to 3 phr sulfur was added. However, when the loading of sulfur was raised to 4 phr, the TBBS requirement increased to 3.5 phr, but the loading of ZnO remained the same at 0.2 phr. In most industrial NR compounds, the loading of sulfur does not exceed 3 phr [2.4,2.6]. Therefore, the 1.5 phr TBBS and 0.3 phr ZnO can be used as a standard recipe in any NR formulation that uses TBBS and ZnO and has less than 4 phr sulfur.

For the BR, the TBBS requirement depended on the loading of sulfur in the cure system. At 0.5 phr sulfur, 1.75 phr TBBS, and at 1 phr sulfur, 3 phr TBBS were needed, respectively, to cause sulfur to react with the rubber. To complete the cure, 0.2 phr ZnO was added to both rubber compounds.

For the EPDM rubber compound with 1 phr sulfur, the TBBS and ZnO requirements to complete the cure were 1 phr and 0.075 phr, respectively. The amount of ZnO required to complete the cure for the NR, BR and EPDM rubber compounds was 0.075-0.3 phr. This is well below the 7 phr that appears in some industrial rubber formulations [2.1]. In addition, the use of one accelerator to cause sulfur to react with the rubber in the absence of ZnO and stearic acid is an interesting property to consider.

Some studies have shown that using fewer chemical curatives had no adverse effects on the mechanical properties of the EPDM, NR and BR rubber vulcanizates. For example, the EPDM vulcanizate that was cured with 1 phr sulfur, 1 phr TBBS and 0.075 phr ZnO, and reinforced with 60 phr mineral kaolin, had a tensile strength of 14.9 MPa, an elongation at break of 1512%, and a tearing energy of 30 kJ/m^2. The NR vulcanizate that was cured with 4 phr sulfur, 3.5 phr TBBS and 0.2 phr ZnO, and reinforced with 60 phr kaolin, had a tensile strength of 22 MPa, an elongation at break of 997%, and a tearing energy of 13 kJ/m^2. The BR vulcanizate that was cured with 0.5 phr sulfur, 1.75 phr TBBS and 0.2 phr ZnO, had a tensile strength of 14.6

MPa, an elongation at break of 889%, and a tearing energy of 7 kJ/m^2 [2.7]. All the indications are that using fewer chemicals is viable, indeed desirable, and improves the cure efficiency.

2.1.7 Summary

Method 1 measured the exact requirements for the TBBS and ZnO to vulcanize the NR, BR and EPDM rubbers at different sulfur loadings.

- For the NR, four cure systems were measured:

(S/TBBS/ZnO): (1/1.5/0.2), (2/1.5/0.3), (3/1.5/0.25), (4/3.5/0.2).

- For the BR, two cure systems were measured:

(S/TBBS/ZnO): (0.5/1.75/0.2), (1/3/0.2).

- For the EPDM, one cure system was measured:

(S/TBBS/ZnO): (1/1/0.075).

Although reducing the chemicals in the cure system to three is a major improvement, the dispersion and distribution of these chemicals in the rubber matrix still pose some difficulties. When TBBS and ZnO are combined into a single additive, it will help to minimize the problem of poor dispersion. A new method (Method 2) will be discussed in the next chapter.

2.2 References

[2.1] – P. A. Ciullo, N. Hewitt, "The rubber formulary", Noyes Publications, NY 1999, p. 79.

[2.2] – http://www.westliberty.edu/health-and-safety/files/2012/08/Stearic-Acid.pdf, Date visited: 11.1.2018.

[2.3] – F. Saeed, A. Ansarifar, R. J. Ellis, Yared Haile-Meskel, "Assessing effect of the re-agglomeration and migration of chemical curatives on the mechanical properties of natural rubber vulcanizate". Adv. Polym. Technol., 32, No. S1, E153-E165 (2013).

[2.4] – R. N. Datta, "Rubber curing systems", Rapra Report 144, 12 (12) (2002), p. 2-37. ISSN:0889-3144

[2.5] – L. Bateman, C. G. Moore, M. Porter, B. Saville, "The chemistry and physics of rubber-like substances", Oxford, MRPRA (1963), 1429.

[2.6] – R. N. Datta, "Rubber curing systems", Rapra Report 144, 12 (12), 2002. ISSN: 0889-3144

[2.7] – S. H. Sheikh, X. Yin, A. Ansarifar, K. Yendall, "The potential of kaolin as a reinforcing filler for rubber composites with new sulfur cure systems". J. Rein Plas & Compos., 36 (16) (2017), 1132-1145.

CHAPTER 3

3 High efficiency sulfur vulcanization of NR with
a single powder
3.1 Method 2 – Use of a single powder to cure NR

3.1.1 Introduction

As demonstrated in the previous chapter, using fewer chemicals to cure rubber does not compromise the efficiency of vulcanization. In fact, it is highly cost-effective and very beneficial to health, safety, and the environment. The next step is to combine the chemicals into a single additive. This is achieved by treating the surface of ZnO with the TBBS and TMTD accelerators in an organic solvent to produce two single powders. The approximate surface areas of these chemicals must be known before this task is undertaken. The quantity of the TBBS and TMTD required to provide monomolecular coverage of the ZnO is calculated based on the approximate surface areas of the ZnO, TBBS, and TMTD molecules used. Gradually, the amounts of the TBBS and TMTD to coat the ZnO are increased to find two materials with optimum properties that can be used to cure the rubber with or without additional sulfur.

3.1.2 Preparation of two single powders and the NR compounds

Single additive 1 - ZnO was treated with TBBS to measure the minimum amount of TBBS needed to satisfactorily crosslink the rubber. Adsorbing the TBBS onto the ZnO provided a convenient single material component to use as an additive. The quantity of TBBS required to provide monomolecular coverage of the ZnO was determined to be 35 mg/g based on the approximate surface areas of the TBBS molecule (6 x 10^{-19} m^2) and the ZnO (50 m^2/g) used. Gradually, the amount of TBBS to coat the ZnO was increased from 35 mg/g to 350 mg/g to find a material with optimum properties. The material with 35 mg/g TBBS led to a very slow cure, but the material with 350 mg/g gave a good cure comparable to a much higher loading of TBBS, which can be as high as 3 phr in some rubber compounds. [3.1] The optimum quantity of TBBS in the powder was 350 mg/g. Hence, 26 wt% of the powder was TBBS and the remaining 74 wt% was ZnO. A large batch was then prepared with this ratio from 202.0 g of ZnO and 70.7 g of TBBS, which was mixed in 100 ml of ethyl acetate or dichloromethane solvent (Sigma Aldrich, UK) in a 500 ml beaker. The suspension was stirred magnetically for 15 min at room temperature (21.5°C). The mixture was filtered under suction using an electric diaphragm vacuum pump (capable of achieving 50 mmHg). The white solid was left to dry overnight and then further dried in a vacuum oven at 50°C. Evaporation of the filtrate on a rotary evaporator showed the mass loss was 0.110 g, indicating the bulk of the TBBS was adsorbed onto the ZnO. The additive will be referred to as the "powder". The powder causes sulfur to react with the rubber to produce stable covalent crosslinks. The temperature of the rubber compounds during mixing was 52-62°C.

Single additive 2 – ZnO was treated with the accelerator by evaporation of a suspension of ZnO in a solution of TMTD in dichloromethane to provide a convenient single material component to use as an additive. The quantity of TMTD required to provide monomolecular coverage of the ZnO was determined to be 24 mg/g based on the approximate surface areas of the TMTD molecule (7.9×10^{-19} m^2) and the ZnO (50 m^2/g) used. Gradually, the amount of TMTD to coat the ZnO was increased from 24 mg/g to 400 mg/g to find a material with optimum properties. The material with 24 mg/g TMTD led to a very slow cure, but the material with 400 mg/g gave a good cure. The optimum quantity of TMTD in the powder was 400 mg/g. Hence, 28.6 wt% of the powder was TMTD and the remaining 71.4 wt% was ZnO. A large batch of the surface-modified ZnO was then prepared with this ratio from 100.0 g of ZnO and 24.0 g of TMTD, which was mixed in 200 ml of the solvent in a 500 ml round-bottomed flask. The suspension was stirred magnetically for 15 minutes at room temperature (21.5°C) to ensure uniform coating. The mixture was evaporated on a rotary evaporator at 100 mbar and further dried at 50°C at 20 mbar for 2 hours to leave a free-flowing white solid. The obtained white solid showed a mass loss of 0.213 g, indicating the bulk of the TMTD was adsorbed onto the ZnO. The sulfur in the TMTD reacts with the rubber to produce crosslinks. The temperature of the rubber compounds during mixing was 52-62°C.

3.1.3 Measurement of the optimum amount of TBBS in the powder for curing NR

The loading of TBBS in the powder was increased progressively from 0.135 phr to 0.383 phr to determine its effects on the cure properties of the rubber with 4 phr sulfur (Table 3.1). This was done at a powder loading of 1.48 phr. The sulfur loading of 4 phr was chosen at random.

Table 3.1: Formulations and cure properties of the rubber compounds with increasing amounts of TBBS in the powder

Formulation (phr)	Compound no									
	1	2	3	4	5	6	7	8	9	10
NR	100	100	100	100	100	100	100	100	100	100
Sulfur	4	4	4	4	4	4	4	4	4	4
TBBS in the powder	0.135	0.192	0.247	0.296	0.342	0.351	0.358	0.367	0.376	0.383
	Curemeter test results at 160°C									
M_L (dNm)	17	15	15	15	15	16	16	16	15	17
M_H (dNm)	32	32	31	31	38	46	45	51	50	53
ΔTorque (dNm)	15	17	16	16	23	30	29	35	35	36
t_{s2} (min)	10.7	8.5	10.4	9.1	4.4	3.2	3.2	3.6	3.5	3.6
t_{95} (min)	55.4	54.5	55.5	54.8	41.4	9.9	10.7	8.4	8.1	8.9
CRI (min⁻¹)	2.2	2.2	2.2	2.2	2.7	14.9	13.3	20.8	21.7	18.9

phr: parts per hundred rubber by weight

When the cure traces of the rubber compounds were examined, the cure for compound 1 (0.135 phr TBBS) rose over a period of 60 min and no plateau was reached (Fig. 3.1a). The curve for compound 6 (0.351 phr TBBS) reached a plateau (Fig. 3.1b). The torque remained at this level for compound 7 (0.358 phr TBBS). Afterwards, a reversion of the vulcanization was observed when the amount of TBBS was increased further to 0.383 phr (compound 10, Fig. 3.1c). A large amount of the TBBS in the powder shortened the cure cycle substantially. Besides, it is interesting that the cure behaviour was so sensitive to small changes in the amount of TBBS in the powder. As Figures 3.1a-3.1c show, the cure curve rose first, then reached equilibrium, and finally, it underwent a reversion as the loading of TBBS in the powder was increased from 24 mg/g to 350 mg/g.

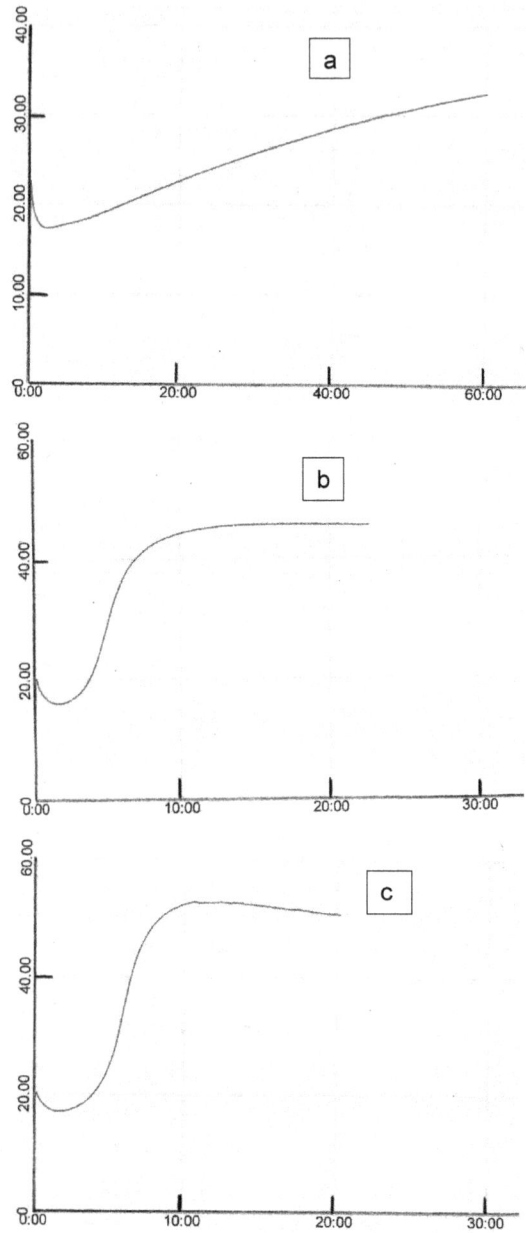

Figures 3.1 – Typical cure traces of the rubber compounds with an increasing loading of TBBS in the powder (Table 3.1), a) Compound 1 (0.135 phr TBBS), b) Compound 6 (0.351 phr TBBS), c) Compound 10 (0.383 phr TBBS).

The increase in the loading of TBBS in the powder had a major influence on the crosslink density and cure rate of the rubber, as indicated by big rises in the values of Δtorque and CRI, respectively. ΔTorque was almost constant at about 15 dNm to 17 dNm with up to 0.296 phr TBBS in the powder. It then rose sharply to 36 dNm when the TBBS loading in the powder reached 0.383 phr (Fig. 3.2).

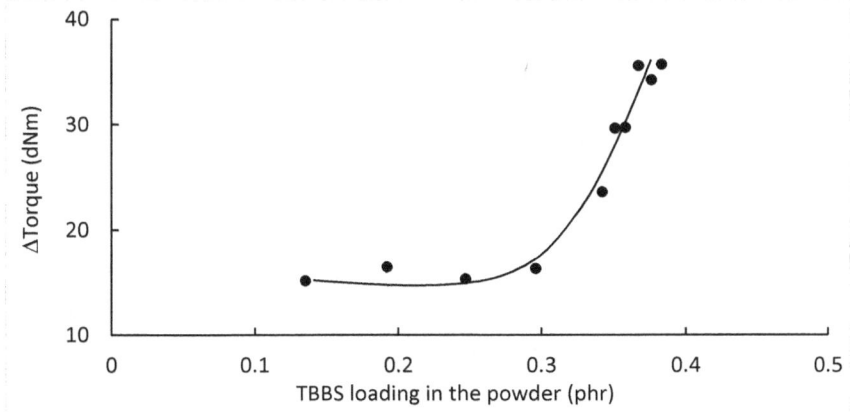

Figure 3.2 - ΔTorque vs. TBBS loading in the powder for the rubber compounds shown in Table 3.1.

The cure rate, as indicated by the CRI, was unaffected by the increase in the TBBS loading in the powder for up to 0.342 phr. But CRI rose sharply to about 21.7 min[-1] when the TBBS loading in the powder reached 0.376 phr. CRI then decreased to 18.9 min[-1] at 0.383 phr TBBS loading in the powder (Fig. 3.3).

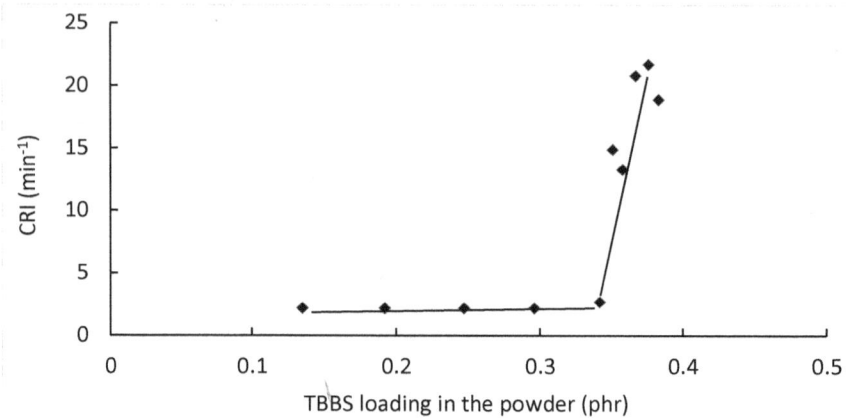

Figure 3.3 - CRI vs. TBBS loading in the powder
for the rubber compounds shown in Table 3.1.

The scorch time was somewhere between 10.7 min and 9.1 min with up to 0.296 phr TBBS in the powder. It then decreased to 4.4 min at 0.342 phr TBBS in the powder. It subsequently reached a plateau around 3.2-3.6 min when the TBBS loading in the powder was raised to 0.383 phr (Fig. 3.4). The increase in the loading of TBBS in the powder for up to 0.296 phr had little or no effect on the optimum cure time, which remained essentially unchanged at about 54.5-55.5 min. However, this was followed by a sharp decrease to 41.4 min at 0.342 phr and then to 10.7 min at 0.358 phr TBBS loading in the powder. The optimum cure time attained a constant value at around 8.1-8.9 min when the loading of TBBS in the powder was increased to 0.383 phr (Fig. 3.4).

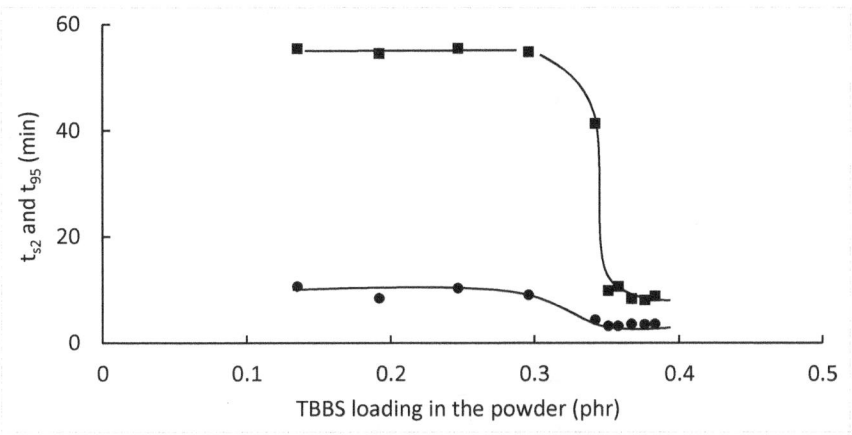

Figure 3.4 - t_{95} and t_{s2} vs. TBBS loading in the powder for the rubber compounds shown in Table 3.1. Optimum cure time (■), scorch time (●).

To summarize, an increase in the amount of TBBS in the powder to above 0.342 phr was greatly beneficial to the cure cycle of the rubber. It reduced the scorch and optimum cure times significantly. The cure rate benefitted at above 0.342 phr TBBS in the powder, with the highest rate, 21.7 min^{-1} recorded at 0.376 phr TBBS in the powder. But the CRI decreased to 18.9 min^{-1} when the TBBS loading in the powder was raised further to 0.383 phr (Table 3.1). The powder with 350 mg/g TBBS (equivalent to 0.383 phr in the formulation) was selected for further work because the rubber compound cured with this powder had the largest Δtorque value, i.e., 36 dNm, and recorded very short scorch and optimum cure times, i.e., 3.6 min and 8.9 min, respectively. For this rubber compound, the CRI was 18.9 min^{-1} (Table 3.1).

3.1.4 Measurement of the optimum amount of the TBBS/ZnO powder for curing NR with sulfur

In these experiments, the amount of TBBS in the powder was kept constant at 350 mg/g, and the loading of the powder in the rubber was increased from 0.63 phr to 5.63 phr at 4 phr sulfur (Table 3.2). The sulfur loading was selected randomly.

Table 3.2: Formulations and cure properties of the rubber compounds with increasing amounts of the TBBS/ZnO powder.

Formulation (phr)	Compound no								
	1	2	3	4	5	6	7	8	9
NR*	100	100	100	100	100	100	100	100	100
Sulfur	4	4	4	4	4	4	4	4	4
Powder	0.63	1.25	1.88	2.5	3.13	3.75	4.38	5	5.63
Curemeter test results at 160°C									
M_L (dNm)	17	16	15	15	15	16	17	16	15
M_H (dNm)	39	46	56	63	68	75	77	80	80
ΔTorque (dNm)	22	30	41	48	53	59	60	64	65
t_{s2} (min)	4.3	3.4	3.3	3.3	3.4	3.6	3.7	3.8	4.0
t_{95} (min)	30.8	7.7	7.0	6.9	6.8	7.2	7.0	7.1	7.4
CRI (min⁻¹)	3.8	23.2	27.0	27.8	29.4	27.8	30.3	30.3	29.4

phr: parts per hundred rubber by weight

When the cure traces were examined for the rubber compound with 0.63 phr powder, the cure reached equilibrium after 40 minutes (Fig. 3.5a). However, when the loading of the powder was raised to above 0.63 phr, the cure underwent reversion soon after 8 minutes, and the cure cycle was a lot shorter. The cure reversion accelerated when the loading of the powder in the rubber kept rising to its highest level, i.e., 5.63 phr (Figs. 3.5b and 3.5c).

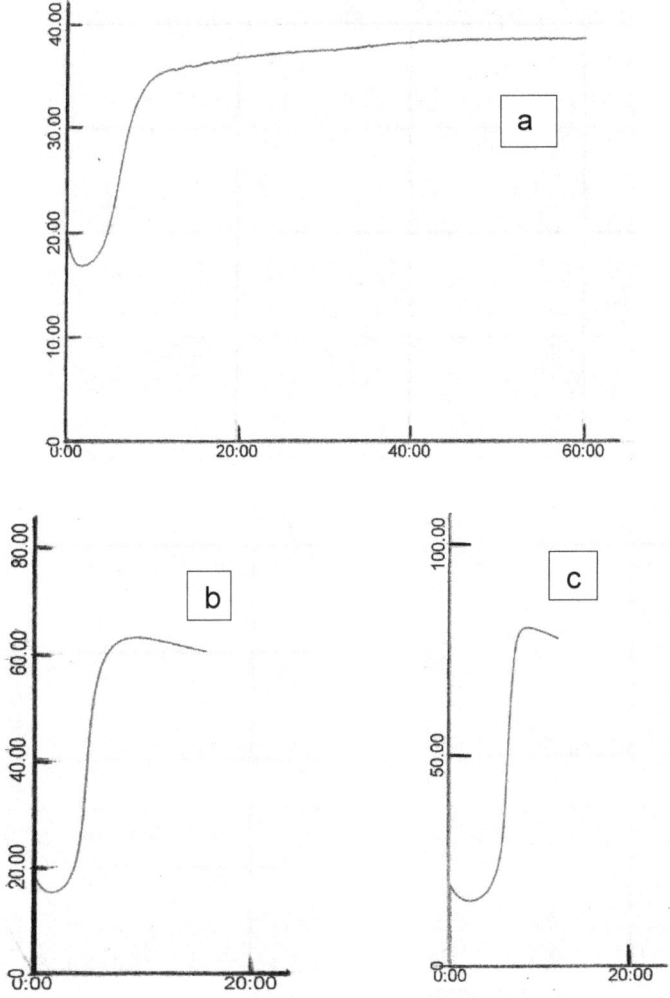

Figure 3.5 - Typical cure traces (Torque (dNm) vs. Time (min)) of the rubber compounds with an increasing loading of the powder. a) compound 1 with 0.63 phr powder; b) compound 4 with 2.5 phr powder; c) compound 9 with 5.63 phr powder.

The minimum torque, M_L, which indicates the uncured rubber viscosity, was not affected by the increase in the loading of the powder in the rubber and remained at around 15 dNm to 17 dNm. The maximum torque, M_H, which showed the extent of crosslinks in the rubber, kept rising from 39 dNm to 80 dNm as the loading of the powder was raised from 0.63 phr to 5.63 phr (Table 3.2). Figure 3.6 shows ∆torque as a function of powder loading. ∆Torque increased from 22 dNm to 48 dNm when the loading of the powder was raised from 0.63 phr to 2.5 phr. It then continued rising at a much slower rate, to about 65 dNm when the loading of the powder reached 5.63 phr. The addition of 2.5 phr powder was enough to cause sulfur to react with the rubber to form stable covalent crosslinks, or chemical bonds, between the rubber chains.

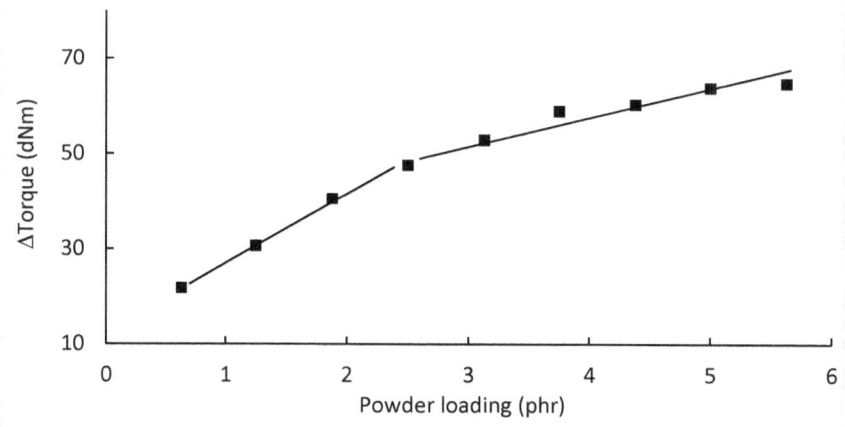

Figure 3.6 - ∆Torque vs. powder loading for the rubber compounds shown in Table 3.2.

Recall that 26 wt% of the powder was TBBS and the remaining 74 wt% was ZnO. On this basis, the 2.5 phr powder contained 0.65 phr TBBS and 1.85 phr ZnO. The scorch time was constant at about 3.3 min to 4.3 min when the full amount of the powder was added. The optimum cure time decreased sharply from 30.8 min to 7.7 min with 1.25 phr powder. It then remained somewhere between 6.8 min and 7.4 min when the loading of the powder was raised to 5.63 phr (Fig. 3.7).

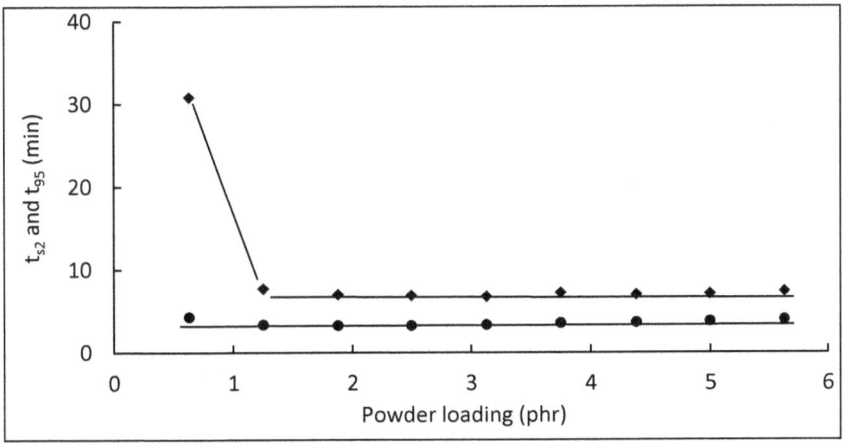

Figure 3.7 - t_{95} and t_{s2} vs. powder loading for the rubber compounds shown in Table 3.2. Optimum cure time (♦), scorch time (●).

The cure rate, as indicated by the CRI, benefited significantly from the addition and progressive increases in the amount of the powder. It rose to 23.2 min^{-1} when 1.25 phr powder was added. The increase was about 510%. Afterwards, it continued rising at a much slower rate, reaching 27.8 min^{-1} when the loading of the powder reached 2.5 phr. It then plateaued at about 27.8 min^{-1} to 30.3 min^{-1} with the full loading of the powder (Fig. 3.8). Above 1.25 phr powder loading, the cure rate did not gain as much, i.e., it increased only by 30%.

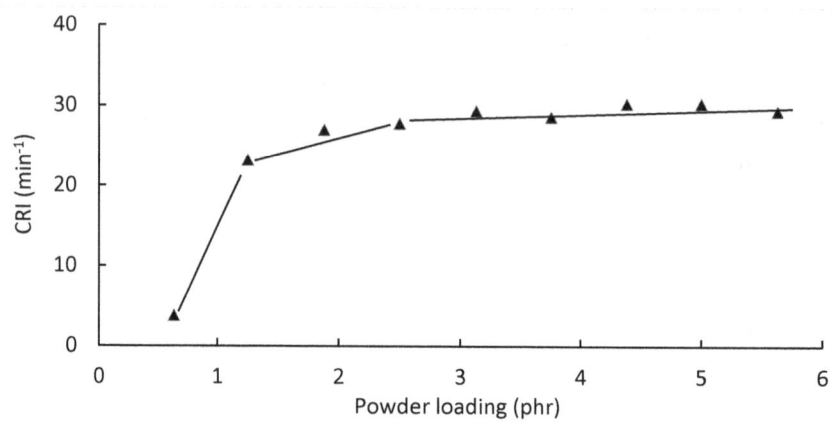

Figure 3.8 - CRI vs. powder loading for the rubber compounds shown in Table 3.2.

For a tire belt skim rubber compound that has 5 phr sulfur and 10.7 phr chemical curatives, the scorch time (t_{s2}) and optimum cure time (t_{90}) are 2.6 min and 9.4 min at 160°C, respectively, and the CRI is 14.7 min^{-1} [3.2]. It is interesting that with 4 phr sulfur and 2.5 phr of the powder made of 0.65 phr TBBS and 1.85 phr ZnO, a shorter optimum cure time and a much higher CRI were measured for rubber compound 4 (Table 3.2) at the same temperature. In fact, a 20 wt% reduction in sulfur and 77 wt% fewer chemical curatives shortened the optimum cure time by 34% (the t_{90} of rubber compound 4 was 6.3 min) and increased the cure rate by over 100%. But the scorch time (t_{s1}) of rubber compound 4 was 27% longer than that of the tire belt skim rubber compound. No stearic acid and no secondary accelerators were used in the cure system of compound 4. The trend observed here suggests that a lower consumption of TBBS and ZnO in sulfur vulcanization yields a significantly shorter cure cycle and hence a more efficient cure. Other benefits include improvements in health, safety, and the environment, as well as a major cost reduction. Treating the surface of ZnO with TBBS to produce a single additive is a more efficient method of using these chemicals in vulcanization than the current methods in use. This has the added advantage of eliminating the secondary accelerators and reducing the amount of ZnO required to complete the cure. Combining the TBBS and ZnO into a single additive by treating the surface of ZnO with the TBBS is undoubtedly the most effective way of minimizing the use of these chemicals and making green compounds for industrial rubber articles.

3.1.5 Effect of an increasing amount of TMTD in the powder on the cure properties of NR

The treatment of ZnO with the sulfur-donor TMTD accelerator in an organic solvent produced an additive that was more efficient than the TBBS/ZnO powder, since TBBS did not have sulfur to donate (Scheme 1). Consequently, no additional sulfur was needed in the vulcanization when the TMTD/ZnO powder was used. To assess the effect of an increasing amount of TMTD in the powder on the cure properties of the rubber, the traces from the curemeter were examined. The loading of the powder was kept constant at 12 phr and the weight ratio of TMTD to ZnO in the powder was increased from 1.09 phr to 3.43 phr to make more powder. The high loading of the powder at 12 phr gave a good cure, whereas lower loadings, e.g., 1.48 phr, produced a poor cure. The tests ran for up to 70 minutes to measure t_{s2}, t_{95}, CRI, M_H, and M_L of the rubber compounds. The results are summarized in Table 3.3.

Table 3.3: Formulations and cure properties of the rubber compounds with an increasing amount of TMTD in the powder. The powder loading was constant at 12 phr.

Formulation (phr)	Compound no						
	1	2	3	4	5	6	7
NR*	100	100	100	100	100	100	100
TMTD/Powder	1.09	1.57	2	2.4	2.77	3.11	3.43
	Curemeter test results at 160°C						
M_L (dNm)	16	17	18	18	18	18	16.5
M_H (dNm)	30	39	43	45	46	50	47
ΔTorque (dNm)	14	22	25	27	28	32	30.5
t_{s2} (min)	3	2.6	2.3	2.3	2.3	2.2	2.5
t_{95} (min)	14.4	14.9	14.1	15.3	14.9	15.7	19
CRI (min^{-1})	8.8	8.1	8.5	7.7	7.9	7.4	6.1

phr: parts per hundred rubber by weight

The cure traces of the rubber compounds in Table 3.3 are presented in Fig. 3.9.

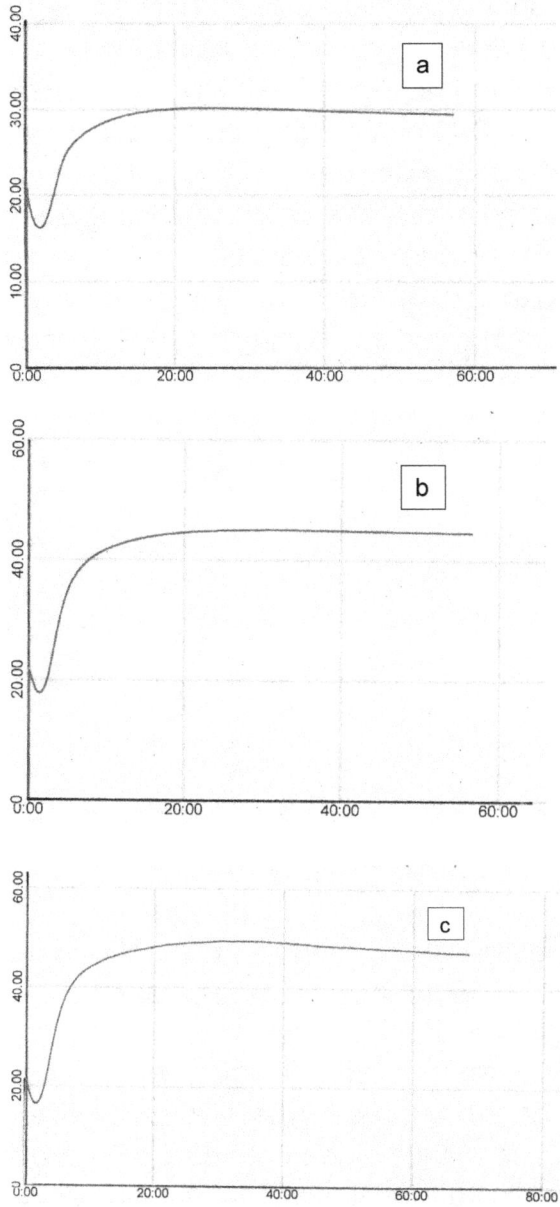

Figure 3.9 – Typical torque (dNm) vs. time (min) traces from ODR tests. Data for the rubber compound with an increasing amount of TMTD in the powder. a) 1.09 phr TMTD in the powder, b) 2.4 phr TMTD in the powder, c) 3.43 phr TMTD in the powder.

Figure 3.10 shows t_{95} and t_{s2} as a function of the loading of TMTD in the powder. The scorch time decreased from 3 min to 2.2 min as the loading of TMTD in the powder was raised from 1.09 phr to 3.11 phr. It then increased to 2.5 min when the TMTD loading in the powder reached 3.43 phr. The optimum cure time was adversely affected by the increase in the loading of TMTD in the powder. It increased from 14.4 min at 1.09 phr TMTD to 15.7 min at 3.11 phr TMTD in the powder. It then rose sharply to 19 min when the loading of TMTD in the powder was raised to 3.43 phr.

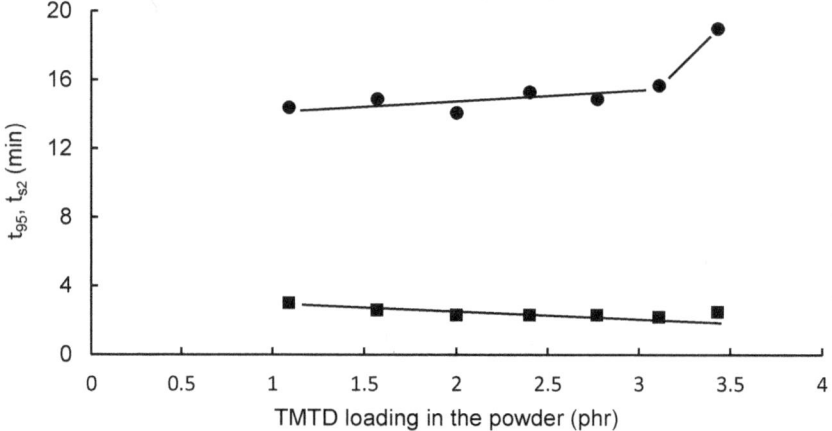

Figure 3.10 - t_{s2} and t_{95} vs. TMTD loading in the powder. (●) t_{95}, (■) t_{s2}.

The addition of an increasing loading of TMTD in the powder was greatly beneficial to the crosslink density, as indicated by a large increase in the value of Δtorque. ΔTorque rose from 14 dNm at 1.09 phr to 22 dNm at 1.57 phr TMTD in the powder. Δtorque then continued rising at a much slower rate, reaching 32.98 dNm when the loading of TMTD in the powder was 3.43 phr (Fig. 3.11).

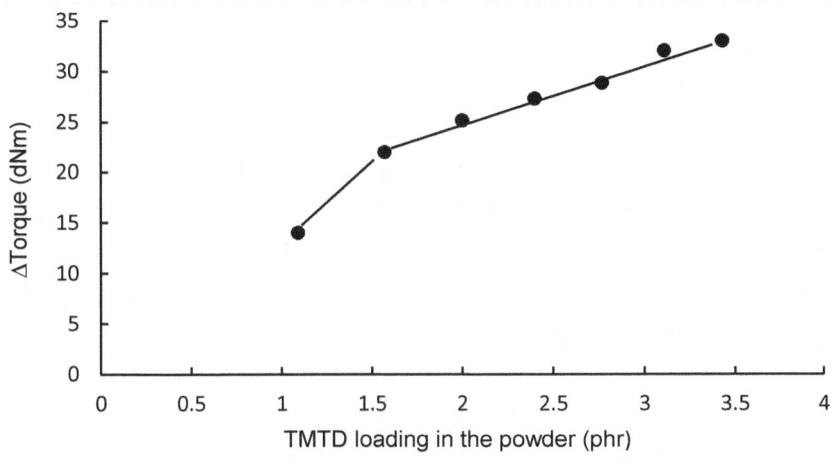

Figure 3.11 - ΔTorque vs. TMTD loading in the powder.

The cure rate, as indicated by CRI, was also adversely influenced by increases in the loading of TMTD in the powder. The CRI decreased from 8.77 min^{-1} at 1.09 phr to 7.41 min^{-1} at 3.11 phr TMTD in the powder. It was subsequently reduced further to 6.07 min^{-1} when the loading of TMTD in the powder reached 3.43 phr (Fig. 3.12).

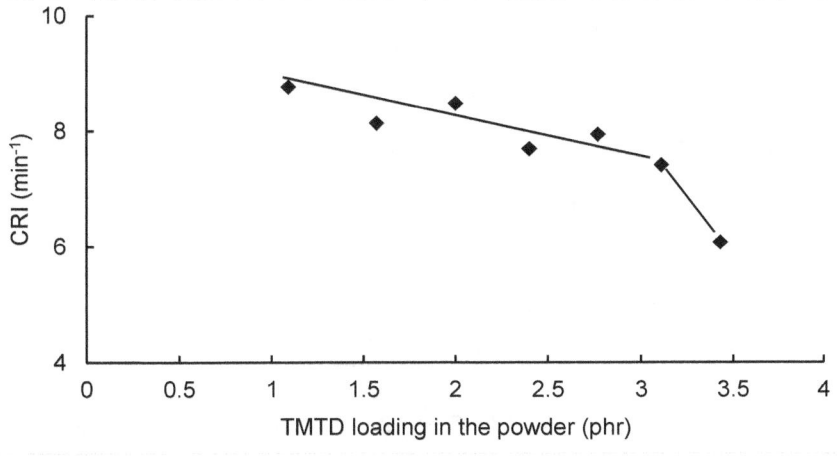

Figure 3.12 - CRI vs. TMTD loading in the powder.

An optimum cure efficiency was achieved by applying multilayers of the TMTD accelerator on the ZnO surface to produce a single material component that was used as an effective additive to cure the rubber. This is contrary to the common practice that uses secondary accelerators in combination with stearic acid, sulfur, and a high loading of ZnO, routinely to improve the efficiency of vulcanization [3.5,3.6]. The effects of an increasing loading of the powder (TMTD/ZnO:400 mg/g) on the cure properties of the NR were then measured.

3.1.6 Effects of an increasing loading of the TMTD/ ZnO powder on the cure properties of NR

After the powder (TMTD/ZnO: 400mg/g) was made, it was mixed with the rubber to produce some compounds. The loading of the powder was increased from 1.48 phr to 20 phr. The t_{s2}, t_{95}, CRI, M_H, and M_L of the rubber compounds were measured. The scorch time decreased from 6.3 min to 2.4 min and the optimum cure time increased from 12 min to 25.9 min as the loading of the powder was raised from 1.48 phr to 20 phr (Fig. 3.13).

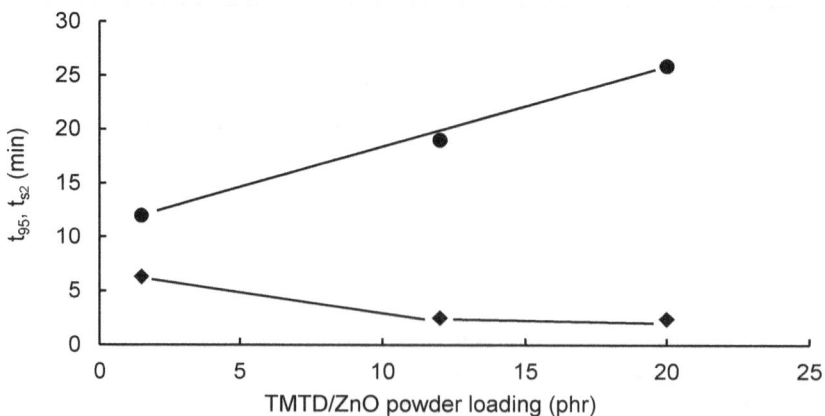

Figure 3.13 - t_{s2} (♦) and t_{95} (●) vs. TMTD/ZnO powder loading.

The addition of an increasing loading of the powder was detrimental to the cure rate. The CRI decreased from 17.54 min^{-1} at 1.48 phr to 4.26 min^{-1} at 20 phr loading of the powder (Fig. 3.14).

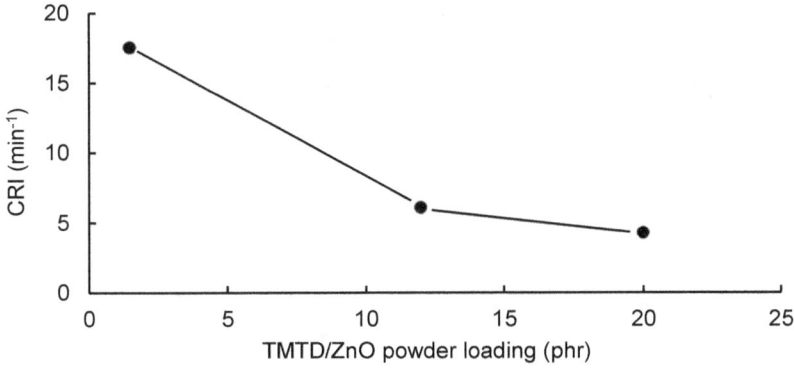

Figure 3.14 - CRI vs. TMTD/ZnO powder loading.

Cure systems and crosslinking density affect the mechanical and dynamic properties of rubber vulcanizates [3.3,3.4]. ΔTorque showed a significant increase from 3.68 dNm to 43.66 dNm as the loading of the powder was raised from 1.48 phr to 20 phr (Fig. 3.15).

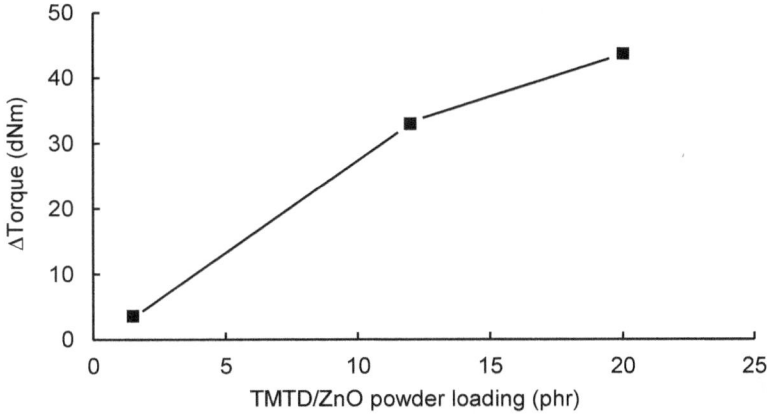

Figure 3.15 - ΔTorque vs. TMTD/ZnO powder loading.

It is interesting that an increase in the crosslink density (as indicated by a rise in the Δtorque) was achieved in the absence of additional sulfur, stearic acid, and secondary accelerators in the cure system. This suggests that a single additive like the sulfur-bearing TMTD/ZnO powder is a very effective chemical for curing the rubber without adding sulfur.

3.1.7 Effect of an increasing loading of the TMTD/ZnO powder on the cure properties of NR with additional sulfur

When the effect of an increasing amount of TMTD in the powder on the cure properties of the rubber was measured, 12 phr of the powder was used to fully cure the rubber (section 3.1.5). This was excessive, and so to reduce the powder requirement, 4 phr sulfur was mixed with the rubber. The effects of 0.63 phr to 5.63 phr of the powder (400mg/g) on the t_{s2}, t_{95}, CRI, M_H, and M_L of the rubber compounds were measured. The tests ran for up to 60 minutes, and the results are summarized in Table 3.4.

Table 3.4: Formulations and cure properties of the rubber compounds with sulfur and an increasing amount of the TMTD/ZnO (400mg/g) powder.

Formulation (phr)	Compound no								
	1	2	3	4	5	6	7	8	9
NR*	100	100	100	100	100	100	100	100	100
Sulfur	4	4	4	4	4	4	4	4	4
Powder	0.63	1.25	1.88	2.5	3.13	3.75	4.38	5	5.63
	Curemeter test results at 160°C								
M_L (dNm)	15	17	17	17	17	18	18	17	17
M_H (dNm)	30	49	59	66	71	74	77	76	74
ΔTorque (dNm)	15	32	42	49	54	56	59	59	57
t_{s2} (min)	2.4	2.2	2.4	2.3	2.2	2.1	2	2.1	2
t_{95} (min)	3.5	4.2	4.7	4.4	4.2	4	3.9	4.1	4.1
CRI (min^{-1})	91	50	43.5	47.6	50	52.6	52.6	50	47.6

phr: parts per hundred rubber by weight

Fig. 3.16 shows some typical cure traces of the rubber compounds listed in Table 3.4.

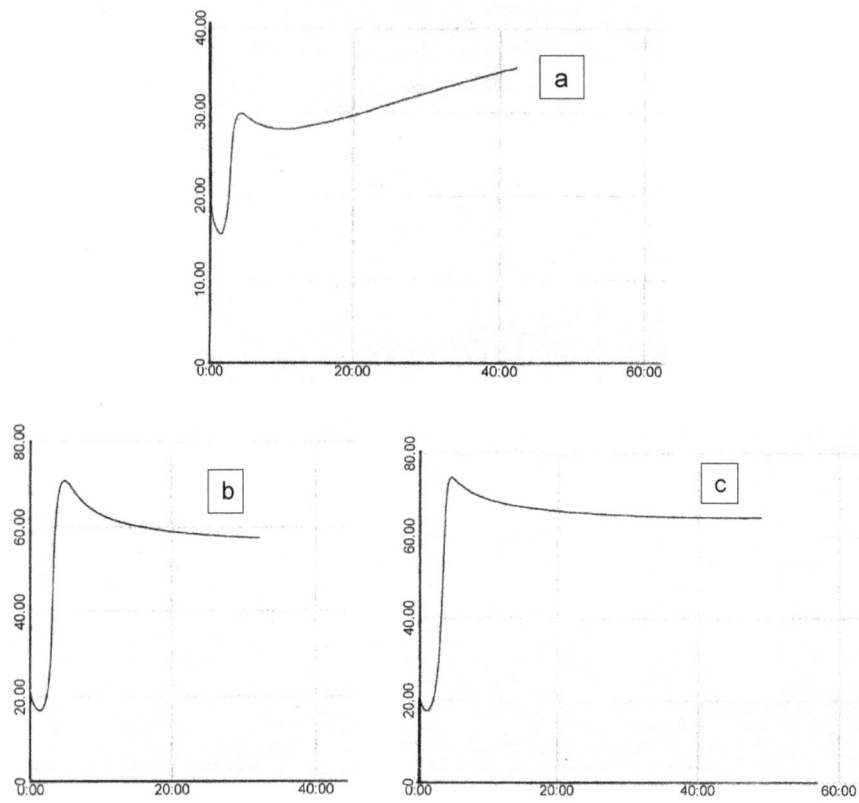

Figure 3.16 – Typical torque (dNm) vs. time (min) traces from ODR tests. Data for the rubber with 4 phr sulfur and the powder. a) 0.63 phr powder, b) 3.13 phr powder, c) 5.63 phr powder.

Figure 3.17 shows Δtorque vs powder loading. The addition of an increasing loading of the powder was very beneficial to the crosslink density, as indicated by a large increase in the value of Δtorque. ΔTorque rose from 15 dNm at 0.63 phr to 49 dNm at 2.5 phr of the powder. ΔTorque continued rising at a slower rate, reaching 57 dNm when the loading of the powder was 5.63 phr. The 2.5 phr of the powder was enough to cause sulfur to react with the rubber to form crosslinks.

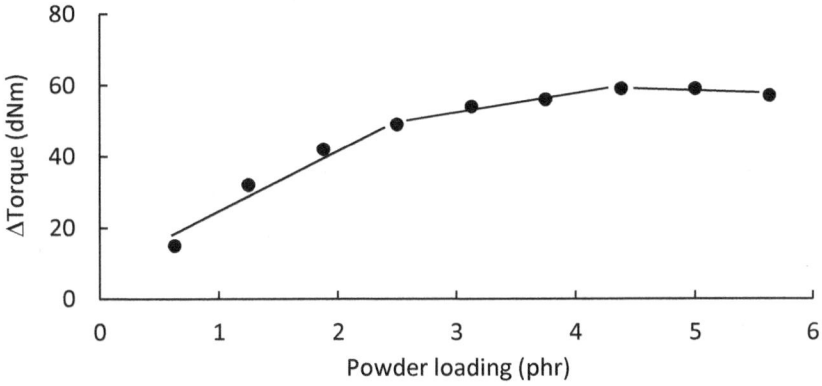

Figure 3.17 - ΔTorque vs. powder loading.

The scorch time decreased from 2.4 min to about 2.0 min when the loading of the powder was increased from 0.63 phr to 5.63 phr. The optimum cure time increased from 3.5 min at 0.63 phr to 4.7 min at 1.88 phr of the powder. It then decreased slightly to about 4.1 min at 5.63 phr of the powder (Fig. 3.18).

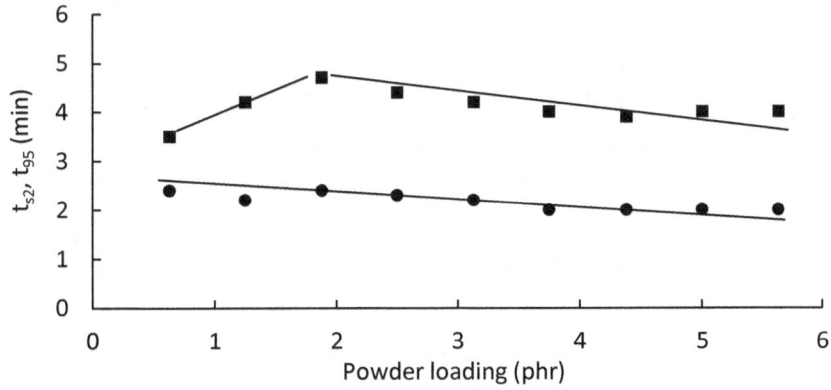

Figure 3.18 - t_{s2} (●) and t_{95} (■) vs. powder loading.

The addition of an increasing loading of the powder was detrimental to the cure rate. The CRI decreased from 91 min^{-1} at 0.63 phr to 50 min^{-1} at 1.25 phr of the powder. It then remained at about 43.5 min^{-1} to 52.6 min^{-1} as the powder loading reached 5.63 phr (Fig. 3.19).

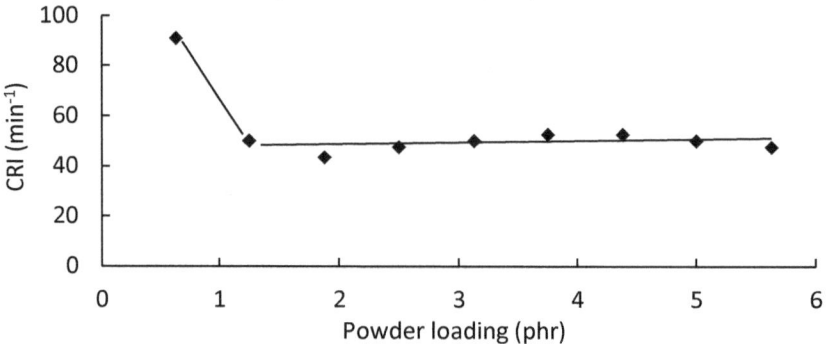

Figure 3.19 - CRI vs. powder loading.

Treating the ZnO with the sulfur-donor TMTD accelerator in an organic solvent provided a single powder that vulcanized the NR efficiently without additional sulfur. However, when 4 phr sulfur was used, the powder requirement for the optimum cure reduced to 2.5 phr. This was 79 wt% less powder. The shortest optimum cure time and the highest CRI were at 3.5 min and 91 min^{-1}, respectively, at 0.63 phr powder loading. Therefore, it is best to use additional sulfur, i.e., 4 phr or less, and a much smaller amount of the powder to secure an efficient cure. The results suggest that when ZnO was treated with the TBBS and TMTD accelerators in an organic solvent to produce two powders, the powders were highly effective in the vulcanization of the rubber with some additional sulfur. The powders can potentially replace the current methods of sulfur vulcanization that are wasteful, expensive, and harmful. But further work will be needed to show the applicability of these methods to other industrially important rubbers such as BR and EPDM. This will be discussed in the next chapter.

3.1.8 Summary

- The NR was cured with 4 phr sulfur and 2.5 phr TBBS/ZnO powder.

- The NR was cured with 12 phr TMTD/ZnO powder.

- The NR was cured with 4 phr sulfur and 2.5 phr TMTD/ZnO powder.

No stearic acid and secondary accelerators were needed to complete the cure.

3.2 References

[3.1] – P. A. Ciullo, N. Hewitt, "The rubber formulary", Noyes Publications, NY 1999, p. 105.

[3.2] – P. A. Ciullo, N. Hewitt, "The rubber formulary", Noyes Publications, NY 1999, p. 79.

[3.3] – B. L. Chan, D. J. Elliott, M. Holley, J. F. Smith, "The influence of curing systems on the properties of natural rubber", J. Polym. Sci., 48, (1974), p. 61-86.

[3.4] – M. Nasir, G. K. Teh, Eur. Polym. J., "The effects of various types of crosslinks on the physical properties of natural rubber", 24 (8), (1988), p. 733-736.

[3.5] – P. A. Ciullo, N. Hewitt, "The rubber formulary", Noyes Publications, NY 1999.

[3.6] – R. N. Datta, Rubber curing systems, Rapra Report 144, 12 (12), 2002

CHAPTER 4

4 High efficiency sulfur vulcanization of BR and
EPDM rubbers with a single powder

4.1 Method 2 – Use of a single powder to cure
BR and EPDM

4.1.1 Effect of an increasing loading of the TBBS/ ZnO powder on the cure properties of BR with sulfur

The powder (TBBS/ZnO: 350mg/g) was used to cure the BR and EPDM rubbers. The loading of the powder in the BR with 4 phr sulfur was raised progressively from 0.63 phr to 5.63 phr to determine its effects on the cure. The sulfur loading of 4 phr was chosen at random. The temperature of the rubber compounds after mixing ended was 55-62°C. For an hour, the rubber compounds were tested in a curemeter to produce traces from which t_{s2}, t_{95}, CRI, M_H, and M_L were measured. The results are summarized in Table 4.1.

Table 4.1: Formulations and cure properties of the BR rubber compounds with the TBBS/ZnO powder and sulfur.

Formulation (phr)	1	2	3	4	5	6	7	8	9
BR	100	100	100	100	100	100	100	100	100
Sulfur	4	4	4	4	4	4	4	4	4
Powder	0.63	1.25	1.88	2.5	3.13	3.75	4.38	5	5.63
Curemeter test results at 160°C									
M_L in dNm	14	14	14	14	15	15	14	14	14
M_H in dNm	51	67	77	83	88	91	93	97	99
ΔTorque in dNm	37	53	63	69	73	76	79	83	85
t_{s2} (min)	9.7	8.6	8.6	8.6	8.4	8.4	7.6	7.5	7.4
t_{95} (min)	58.3	48	38.6	22.4	22.8	23.9	21.5	21.8	22.7
CRI (min^{-1})	2.1	2.5	3.3	7.3	6.9	6.5	7.2	7.0	6.5

phr: parts per hundred rubber by weight

Figure 4.1 shows t_{s2} and t_{95} as a function of the powder loading. The scorch time was unchanged at about 7.4 min to 9.7 min, as the loading of the powder was increased from 0.63 phr to 5.63 phr. The optimum cure time decreased sharply from 58.3 min at 0.63 phr to 22.4 min at 2.5 phr of the powder. It then remained at a steady value of about 21.5 min to 23.0 min when the loading of the powder reached 5.63 phr.

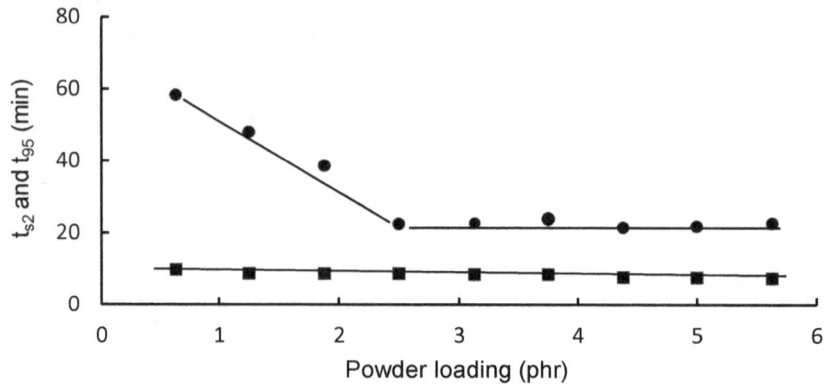

Figure 4.1 - t_{95}(●) and t_{s2}(■) vs. powder loading for the rubber compounds shown in Table 4.1.

The cure rate, as indicated by the CRI, rose from 2.1 min^{-1} at 0.63 phr to 3.3 min^{-1} at 1.88 phr of the powder. It then increased to 7.3 min^{-1} at 2.5 phr of the powder. The CRI started to decrease slowly to about 6.5 min^{-1} as the loading of the powder reached 5.63 phr (Fig. 4.2).

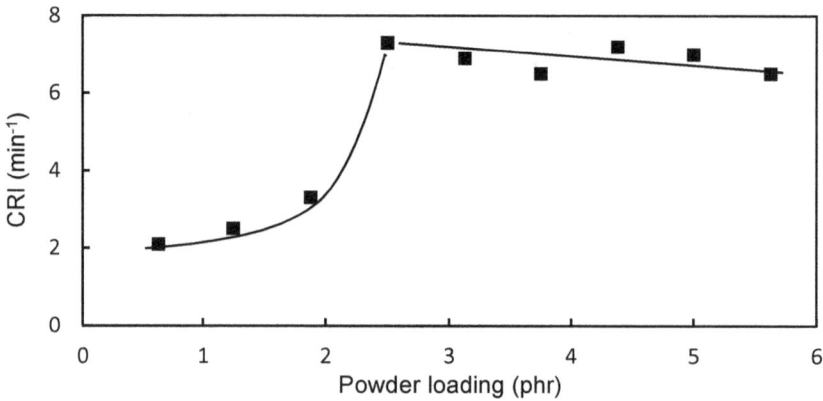

Figure 4.2 – CRI vs. powder loading for the rubber compounds shown in Table 4.1.

For these rubber compounds, Δtorque increased sharply from 37 dNm to 69 dNm when the loading of the powder was raised from 0.63 phr to 2.5 phr. Δtorque then rose at a much slower rate, reaching 85 dNm, when the full amount of the powder, i.e., 5.63 phr, was added to the rubber (Fig. 4.3). The 2.5 phr of the powder was enough to fully cure the rubber.

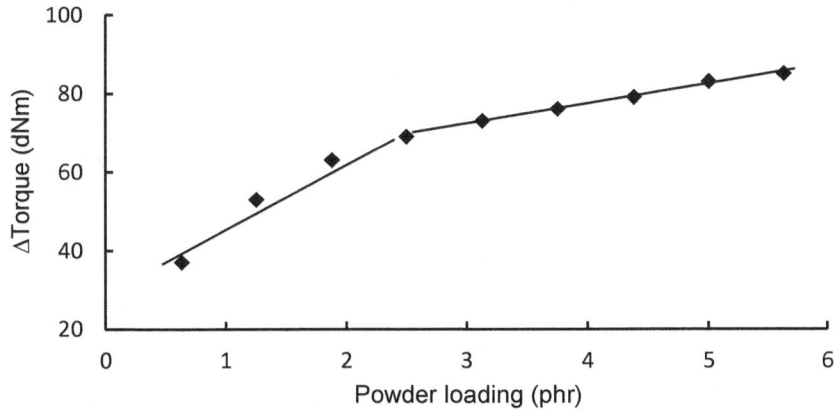

Figure 4.3 - ΔTorque vs. powder loading for the rubber compounds shown in Table 4.1

The cure traces of the rubber compounds in Table 4.1 are shown in Fig. 4.4. The cure was marching for compounds 1- 3. It reached an equilibrium for compounds 4 and 5. And it exhibited a reversion for compounds 6-9.

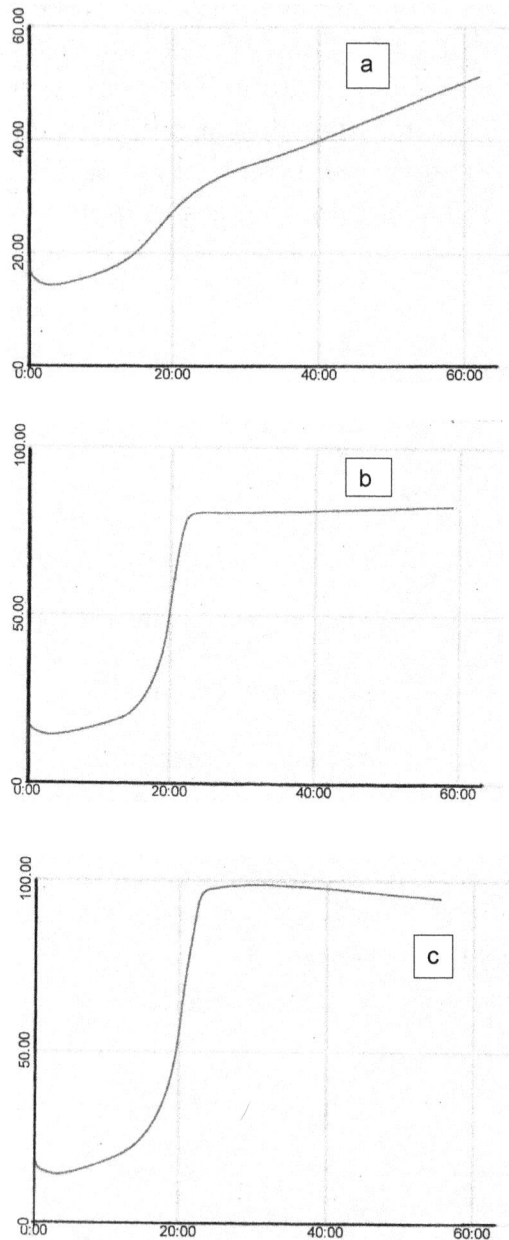

Figure 4.4 - Typical torque (dNm) vs. time (min) cure traces of the BR rubber compounds. a) Compound 1 (0.63 phr powder), b) Compound 4 (2.5 phr powder), c) Compound 9 (5.63 phr powder).

The sulfur cure system in the BR-based Press Molded Closed Cell Sheet Sponge is made of two accelerators and two activators, which add up to 1.4 phr and 11 phr, respectively [4.1]. In total, 12.4 phr chemicals are used to fully cure the article with 4 phr sulfur. As Figure 4.3 shows, the optimum loading of the powder for curing the BR rubber compound with 4 phr sulfur was 2.5 phr. This was nearly 80 wt% lower than the chemicals used for curing the article mentioned above. Therefore, the powder, in combination with sulfur is a more efficient method for vulcanizing the article.

4.1.2 Effect of an increasing loading of the TBBS/ ZnO powder on the cure properties of EPDM with 1 phr sulfur

The powder (TBBS/ZnO: 350mg/g) was used to cure the EPDM rubber with sulfur. The loading of the powder in the rubber with 1 phr sulfur was raised progressively from 0.75 phr to 5.5 phr to determine its effects on the cure. The temperature of the rubber compounds after mixing ended was 55-62°C. The rubber compounds were tested in a curemeter for 2 hours to produce traces from which t_{s2}, t_{95}, CRI, M_H, and M_L were measured (Table 4.2).

Table 4.2: Formulations and cure properties of the EPDM rubber compounds with an increasing loading of the TBBS/ZnO powder and sulfur.

Formulation (phr)	1	2	3	4	5	6	7	8	9
EPDM	100	100	100	100	100	100	100	100	100
Sulfur	1	1	1	1	1	1	1	1	1
Powder	0.75	1.0	1.7	2.5	3.2	4	4.5	5	5.5
Curemeter test results at 160°C									
M_L (dNm)	15	15	16	14	15	14	15	14	15
M_H (dNm)	61	64	77	84	92	89	96	92	96
ΔTorque (dNm)	46	49	61	70	77	75	81	78	81
t_{s2} (min)	21.2	23	22.2	19	18.1	20	20.2	22.7	21.5
t_{95} (min)	98.2	93.7	89.3	78	71.4	67.3	72.7	69.6	68.5
CRI (min⁻¹)	1.3	1.4	1.5	1.7	1.9	2.1	1.9	2.1	2.1

phr: parts per hundred rubber by weight

Some typical cure traces from the rubber compounds listed in Table 4.2 are shown in Figs. 4.5a, 4.5b, and 4.5c. The cure was marching for all the rubber compounds tested.

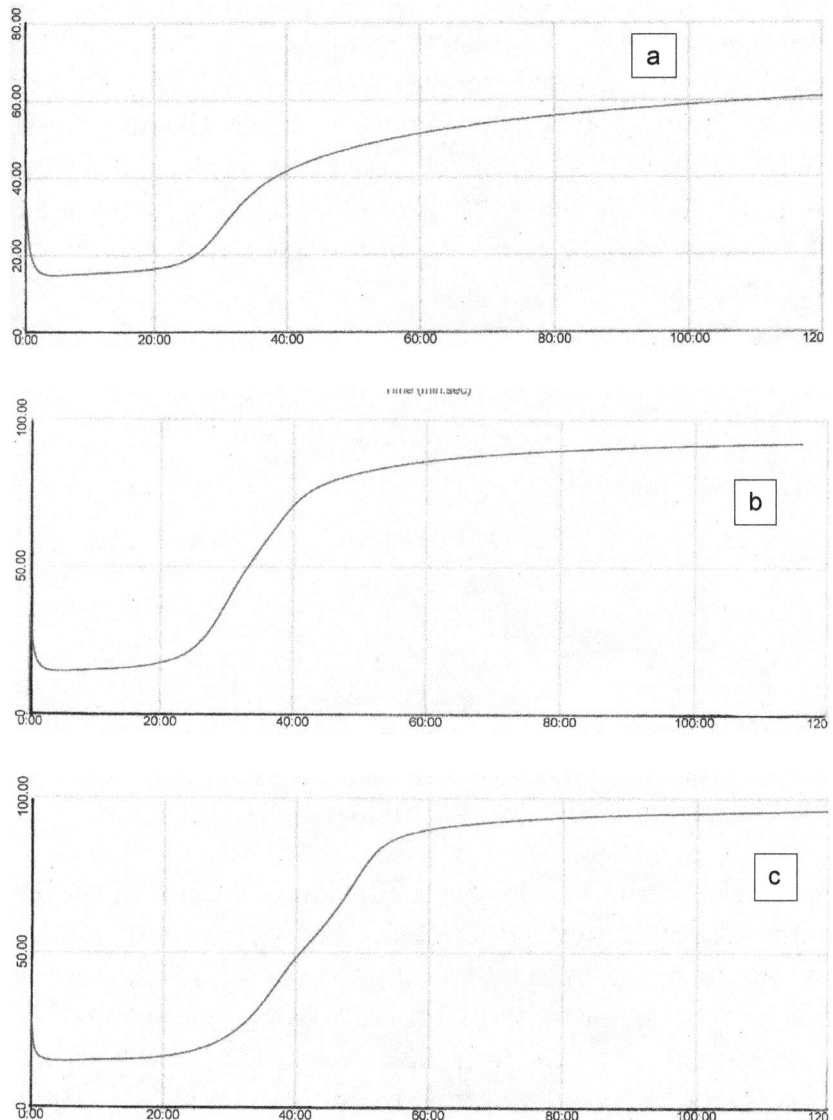

Figure 4.5 – Typical torque (dNm) vs. time (min) cure traces of the EPDM rubber compounds with 1 phr sulfur and the powder. a) Compound 1 (0.75 phr powder), b) Compound 5 (3.2 phr powder), c) Compound 9 (5.5 phr powder).

Figure 4.6 shows Δtorque vs powder loading for the rubber compounds tested. ΔTorque increased from 46 dNm to 77 dNm when the loading of the powder was raised from 0.75 phr to 3.2 phr. ΔTorque then continued rising at a much slower rate to about 81 dNm when the loading of the powder reached 5.5 phr. The addition of 3.2 phr of the powder was enough to cause the sulfur to react with the rubber to form stable covalent crosslinks, or chemical bonds, between the chains.

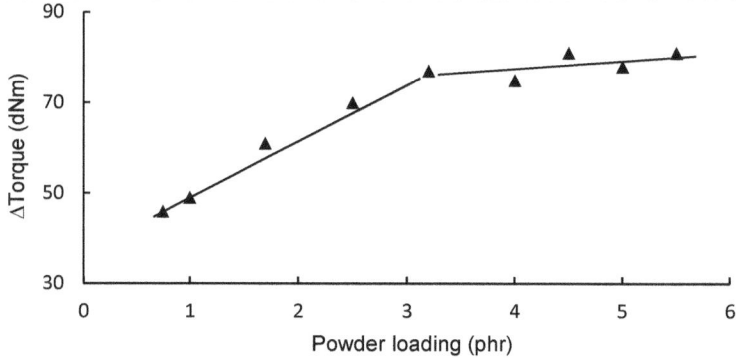

Figure 4.6 - ΔTorque vs. powder loading for the rubber compounds shown in Table 4.2.

Figure 4.7 shows t_{s2} and t_{95} as a function of the powder loading for the rubber compounds tested. The scorch time remained at about 18.1 min to 23 min as the loading of the powder was raised progressively from 0.75 phr to 5.5 phr. The optimum cure time decreased sharply from 98.2 min at 0.75 phr to 71.4 phr at 3.2 phr of the powder. It then remained at a steady value of 68.5-72.7 min when the loading of the powder reached 5.5 phr.

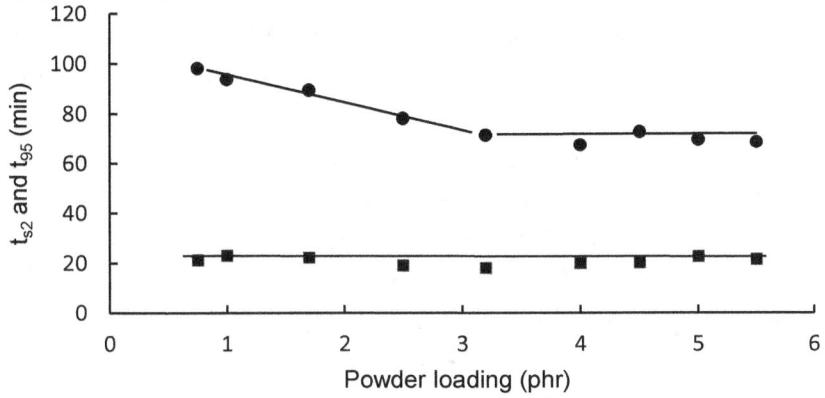

Figure 4.7 - t_{95} (●) and t_{s2} (■) vs. powder loading for the rubber compounds shown in Table 4.2.

The cure rate, as indicated by the CRI, rose from 1.3 min^{-1} at 0.75 phr to 2.1 min^{-1} at 4 phr of the powder. It subsequently plateaued at about 1.9-2.1 min^{-1} when the full amount of the powder was incorporated into the rubber (Fig. 4.8). The CRI result of 1.9 min^{-1} at 4.5 phr of the powder is an anomaly.

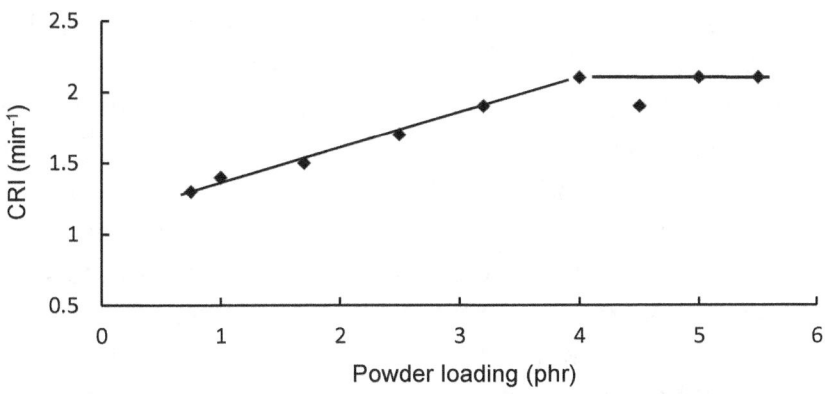

Figure 4.8 – CRI vs. powder loading for the rubber compounds shown in Table 4.2.

When the powder loading exceeded 4 phr, the rubber was brittle and weak. The composition of the powder is 26 wt% TBBS and 74 wt% ZnO. On this basis, 3.2 phr of the powder is made of 0.83 phr TBBS and 2.37 phr ZnO. The sulfur cure system in the EPDM-based Curtain Wall Seal has two accelerators, adding up to 2.75 phr, and two activators (ZnO and stearic acid), adding up to 6 phr [4.2]. In total, 8.75 phr chemicals are used to cure the article with 1 phr sulfur. The optimum loading of the powder for curing the EPDM rubber with 1 phr sulfur was 3.2 phr (Fig. 4.6). This was 63 wt% lower than the chemicals used for curing the Curtain Wall Seal. Therefore, the powder with added sulfur is more efficient for vulcanizing the rubber. No stearic acid was used in the cure system with the powder.

4.1.3 Effect of an increasing loading of the TBBS/ ZnO powder on the cure properties of EPDM with different amounts of sulfur

The powder (TBBS/ZnO:350mg/g) was used to cure the EPDM rubber with different amounts of sulfur. The loading of the powder was raised progressively from 0.75 phr to 7 phr and that of sulfur from 1 phr to 4 phr to determine their effects on the cure properties of the rubber. The temperature of the rubber compounds after mixing ended was 57-65°C. The rubber compounds were tested in a curemeter to produce traces from which t_{s2}, t_{95}, CRI, M_H, and M_L were measured (Table 4.3). Despite the tests lasting almost two hours, the results summarized in Tables 4.3-4.6 correspond to the first 60 minutes of the cure. The cure properties listed in Table 4.3 were measured from the cure traces, some of which are shown in Fig 4.5.

Table 4.3: Formulations and cure properties of the EPDM rubber compounds with an increasing loading of the TBBS/ZnO powder and 1 phr sulfur. Data taken from Fig. 4.5

Formulation (phr)	1	2	3	4	5	6	7	8	9
EPDM	100	100	100	100	100	100	100	100	100
Sulfur	1	1	1	1	1	1	1	1	1
TBBS/ZnO Powder	0.75	1.0	1.7	2.5	3.2	4	4.5	5	5.5
Curemeter test results at 160°C									
M_L (dNm)	15	15	16	19	15	14	15	14	15
M_H (dNm)	51.5	55	67	53	85	84	88	85	89.5
ΔTorque (dNm)	36.5	40	51	62	70	70	73	71	74.5
t_{s2} (min)	21	23	22	19	18	20	20	23	21.5
t_{95} (min)	56	54	56	53	71	67	54	52	54
CRI (min^{-1})	2.9	3.2	2.9	2.9	1.9	2.1	2.9	2.1	3.1

Table 4.4: Formulations and cure properties of the EPDM rubber compounds with the powder and 2 phr sulfur.

Formulation (phr)	1	2	3	4	5	6	7	8	9
EPDM	100	100	100	100	100	100	100	100	100
Sulfur	2	2	2	2	2	2	2	2	2
TBBS/ZnO Powder	1.5	2.5	3.13	3.75	4.38	5	5.63	6	6.3
Curemeter test results at 160°C									
M_L (dNm)	14	15	14	15	14	15	15	15	15
M_H (dNm)	66	84	87	98	96.5	105	108	108	108
ΔTorque (dNm)	52	69	73	83	82.5	90	93	93	93
t_{s2} (min)	15	15	15	15	16	16	16	17	17
t_{95} (min)	48	52	51	49	52	48	48	49	49
CRI (min^{-1})	3	2.7	2.8	2.9	2.8	3.1	3.1	3.1	3.1

Some typical cure traces for the rubber compounds listed in Table 4.4 are shown in Fig. 4.9.

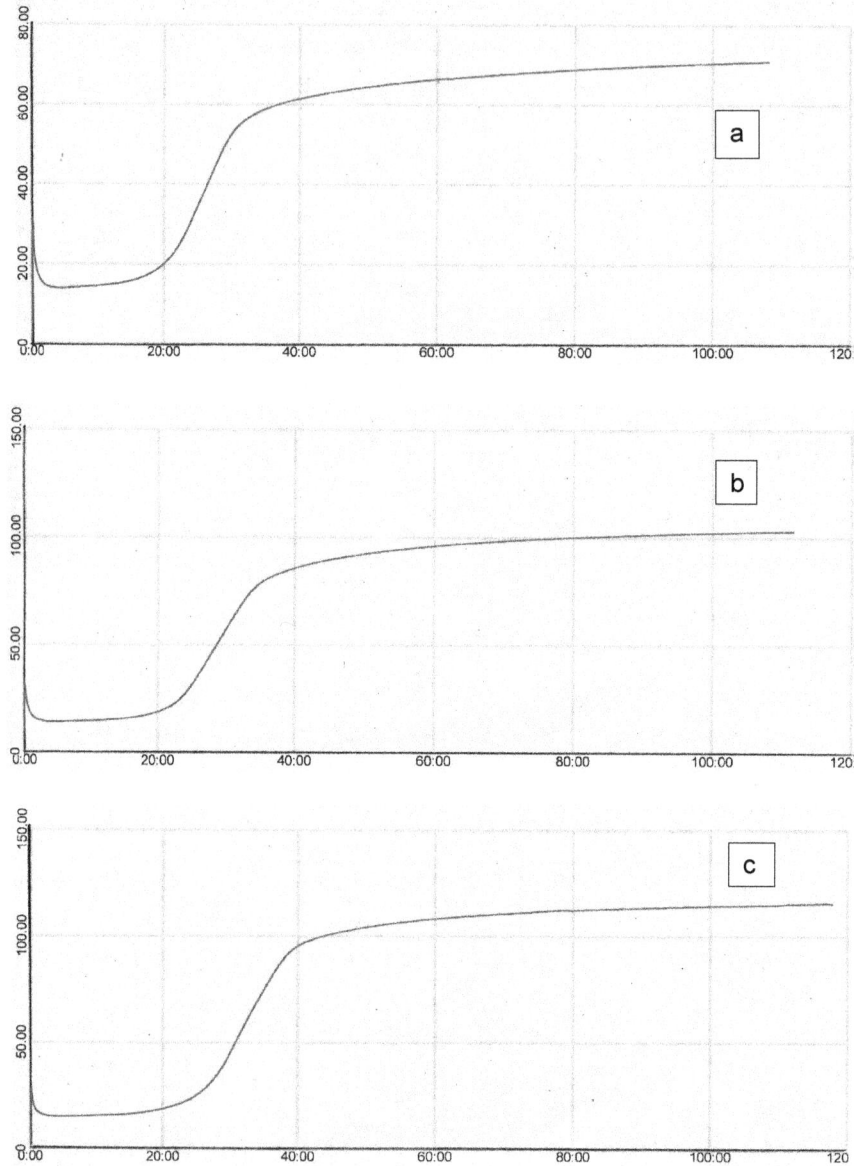

Figure 4.9 – Typical torque (dNm) vs. time (min) cure traces of the EPDM rubber compounds with 2 phr sulfur and the powder. a) Compound 1 (1.5 phr powder), b) Compound 5 (4.38 phr powder), c) Compound 9 (6.3 phr powder).

Table 4.5: Formulations and cure properties of the EPDM rubber compounds with the powder and 3 phr sulfur.

Formulation (phr)	1	2	3	4	5	6	7	8	9
EPDM	100	100	100	100	100	100	100	100	100
Sulfur	3	3	3	3	3	3	3	3	3
TBBS/ZnO Powder	2.5	3.13	3.75	4.38	5	5.63	6	6.3	7
Curemeter test results at 160°C									
M_L (dNm)	16	15	15	16	15	16	15	15	15
M_H (dNm)	88	95	99	100	109	109	111	114	118
ΔTorque (dNm)	72	80	84	84	94	93	96	99	103
t_{s2} (min)	10	11	11	14	11	12	11	12	13
t_{95} (min)	41	43	44	46	44	46	42	42	45
CRI (min^{-1})	3.2	3.1	3	3.1	3	2.9	3.3	3.3	3.1

Some typical cure traces for the rubber compounds listed in Table 4.5 are shown in Fig. 4.10.

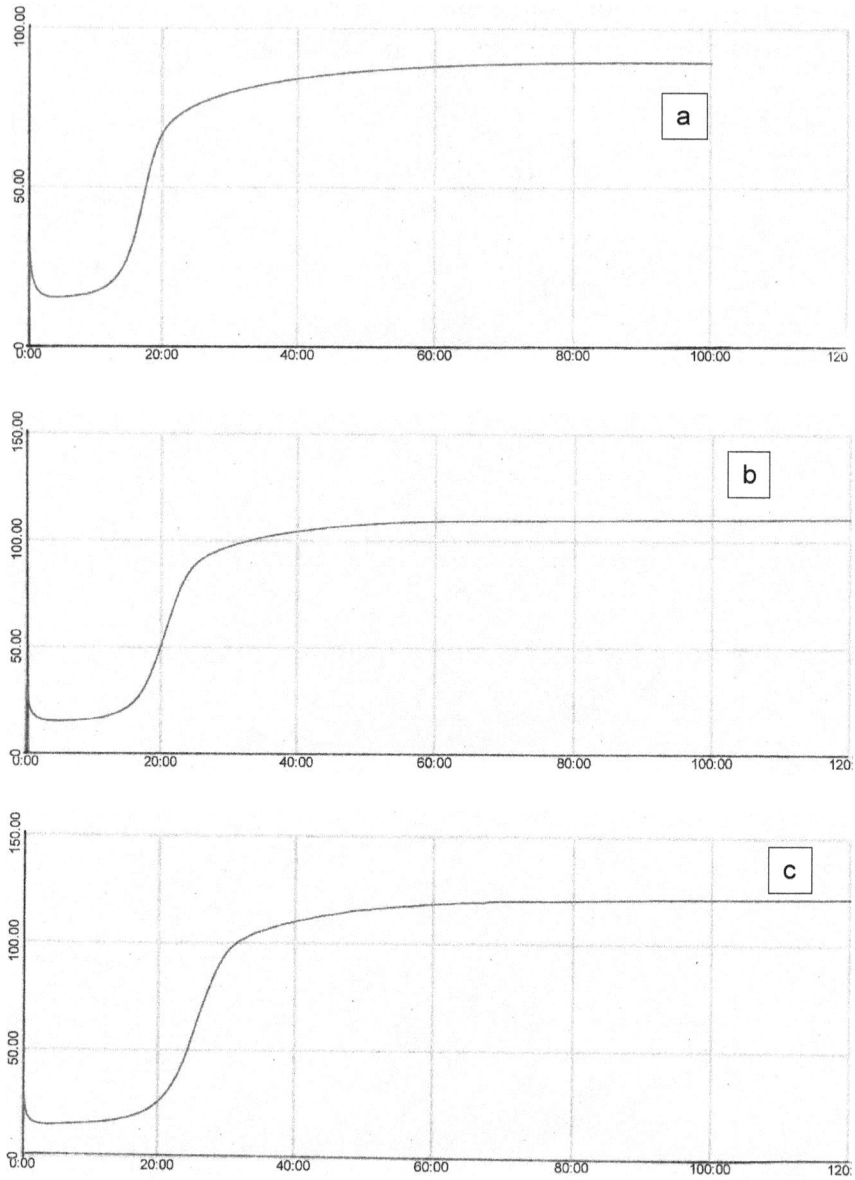

Figure 4.10 – Typical torque (dNm) vs. time (min) cure traces of the EPDM rubber compounds with 3 phr sulfur and the powder. a) Compound 1 (2.5 phr powder), b) Compound 5 (5 phr powder), c) Compound 9 (7 phr powder).

Table 4.6: Formulations and cure properties of the EPDM rubber compounds with the powder and 4 phr sulfur.

Formulation (phr)	1	2	3	4	5	6	7	8	9
EPDM	100	100	100	100	100	100	100	100	100
Sulfur	4	4	4	4	4	4	4	4	4
Powder	2.5	3.13	3.75	4.38	5	5.63	6	6.3	7
Curemeter test results at 160°C									
M_L (dNm)	15	15	15	15	15	15	14	14	15
M_H (dNm)	84	92	98	103	107	113	114	115	120
ΔTorque (dNm)	69	77	83	88	92	98	100	101	105
t_{s2} (min)	9	10	10	10	10	11	10	11	12
t_{95} (min)	35	37	39	39	39	42	40	41	41
CRI (min^{-1})	3.9	3.7	3.5	3.5	3.5	3.2	3.3	3.3	3.4

Some typical cure traces for the rubber compounds listed in Table 4.6 are shown in Fig. 4.11

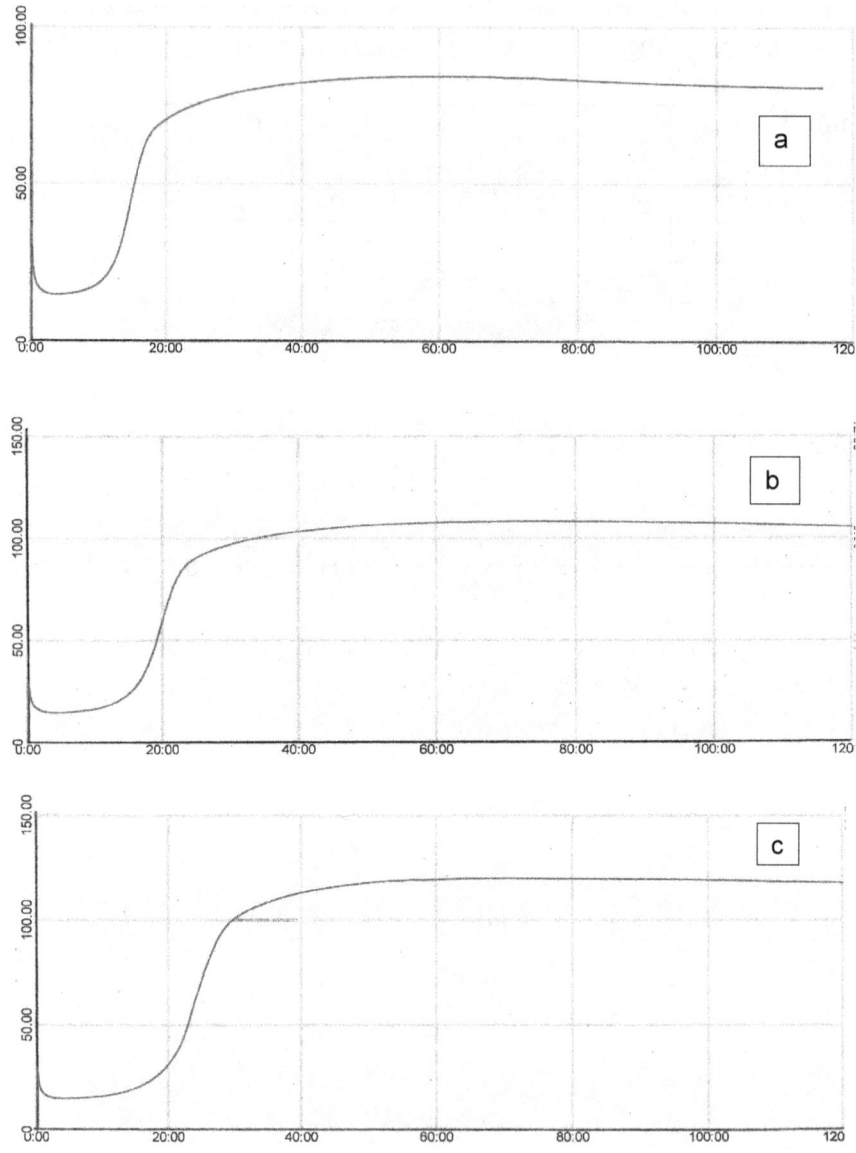

Figure 4.11– Typical torque (dNm) vs. time (min) cure traces of the EPDM rubber compounds with 4 phr sulfur and the powder. a) Compound 1 (2.5 phr powder), b) Compound 5 (5 phr powder), c) Compound 9 (7 phr powder).

The cure for the rubber compounds with 1 phr to 3 phr sulfur was marching (Figs. 4.5, 4.9, 4.10), but for the rubber compound with 4 phr sulfur, it reached equilibrium after 60 minutes (Fig. 4.11).

For the rubber compound with 1 phr sulfur, Δtorque increased from 36.7 dNm at 0.75 phr to 73.5 dNm at 3.2 phr of the powder. It then continued rising at a very slow rate, reaching 74.5 dNm when the loading of the powder was increased to 5.5 phr. The 3.2 phr of the powder was enough to react the sulfur with the rubber and cure the rubber compound fully (Fig. 4.12).

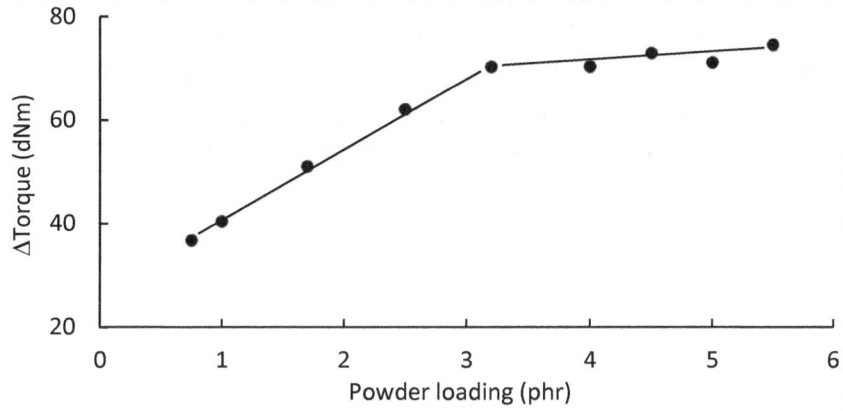

Figure 4.12 - ΔTorque vs. powder loading for
the rubber compound with 1 phr sulfur.

For the rubber compound with 2 phr sulfur, Δtorque rose from 52 dNm to 82.7 dNm when the loading of the powder was increased from 1.5 phr to 3.75 phr. The rate of increase in Δtorque slowed down, reaching 92.8 dNm, as the loading of the powder was raised to 6.3 phr. At this loading of sulfur, 3.75 phr was the optimum powder loading for curing the rubber (Fig. 4.13).

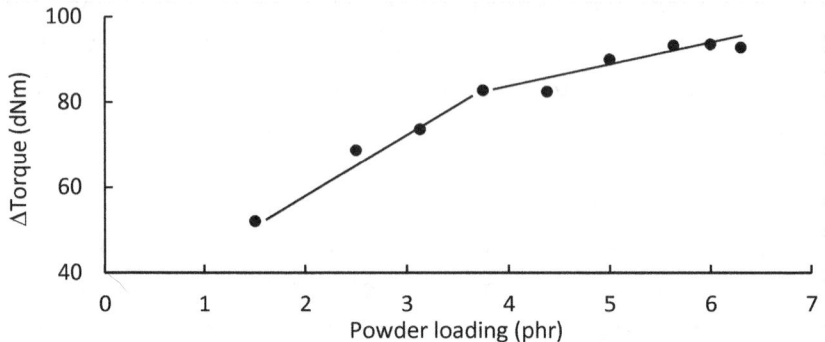

Figure 4.13 - ΔTorque vs. powder loading for
the rubber compound with 2 phr sulfur.

For the rubber compound with 3 phr, Δtorque was 93.8 dNm
at 5 phr of the powder. It then continued rising at a slower rate,
reaching 102.3 dNm at 7 phr of the powder. For this compound,
the optimum powder loading was 5 phr (Fig. 4.14).

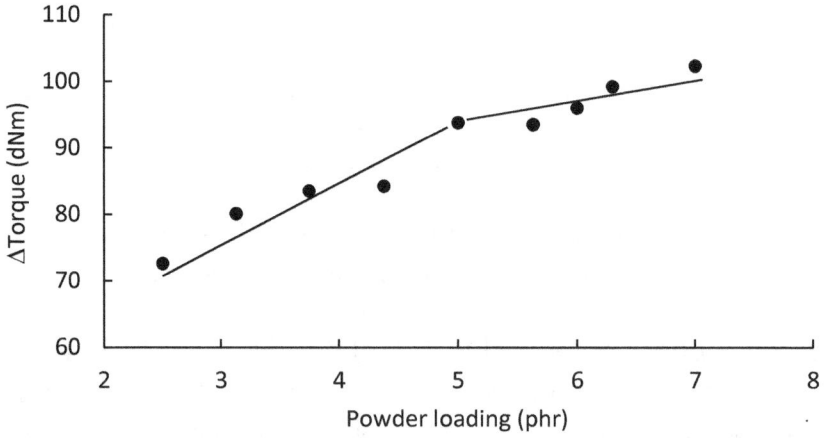

Figure 4.14 - ΔTorque vs. powder loading for
the rubber compound with 3 phr sulfur.

For the rubber compound with 4 phr sulfur, Δtorque kept
rising until it reached 98.2 dNm at 5.63 phr of the powder. It
then rose to 104.7 dNm when the full amount of the powder,

i.e., 7 phr, was incorporated into the rubber (Fig. 4.15). For this rubber compound, the optimum powder loading was at 5.63 phr.

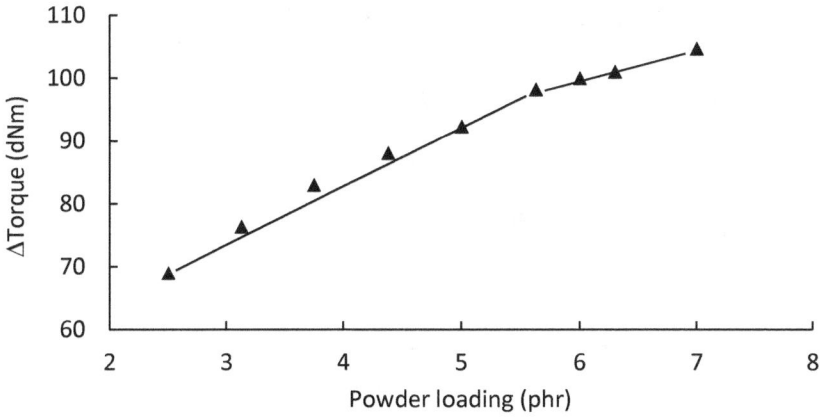

Figure 4.15 - ΔTorque vs. powder loading
for the compound with 4 phr sulfur.

The correlation between the optimum powder loading and the sulfur loading was linear. The powder requirement for a full cure increased as a function of the sulfur loading (Fig. 4.16).

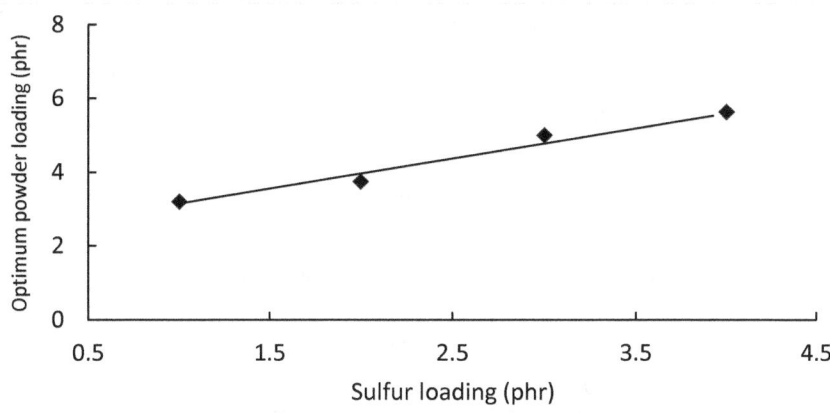

Figure 4.16 - Optimum powder loading vs. sulfur loading.

Figures 4.17- 4.20 show t_{s2} and t_{95} vs. powder loading for the rubber compounds cured with 1 phr, 2 phr, 3 phr and 4 phr sulfur. For the 1 phr, 2 phr, and 3 phr loadings of sulfur, both the scorch and optimum cure times remained constant as the powder loading was increased from 0.75 phr to 7 phr.

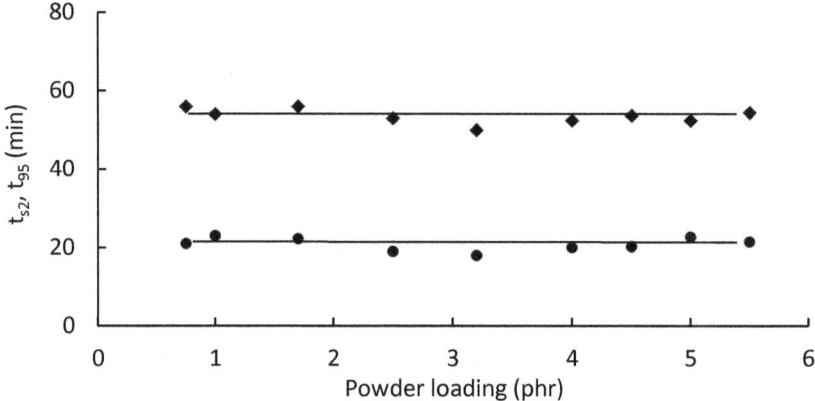

Figure 4.17 - t_{s2} (●) and t_{95} (♦) vs. powder loading at 1 phr sulfur loading.

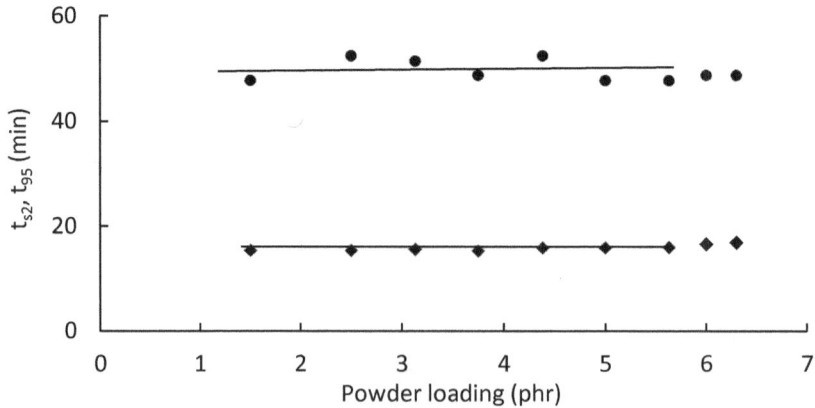

Figure 4.18 - t_{s2} (♦) and t_{95} (●) vs. powder loading at 2 phr sulfur loading.

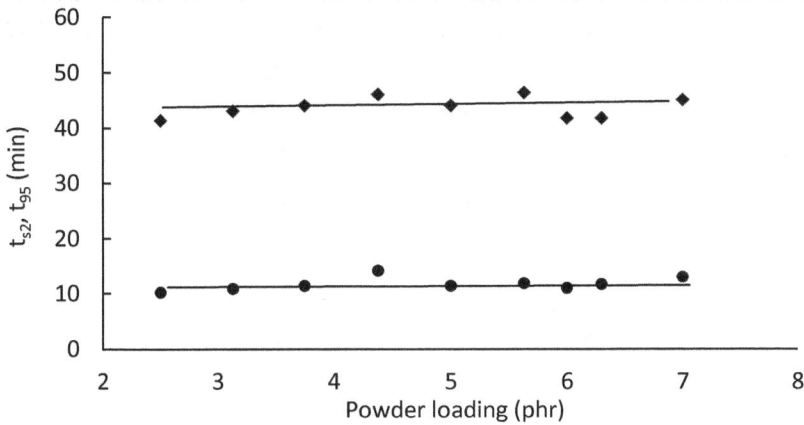

Figure 4.19 - t_{s2} (●) and t_{95} (♦) vs. powder
loading at 3 phr sulfur loading.

For the rubber compound cured with 4 phr sulfur, both the
scorch and optimum cure times increased as a function of the
powder loading, with the scorch time rising by 36% and the
optimum cure time by 20% (Fig. 4.20).

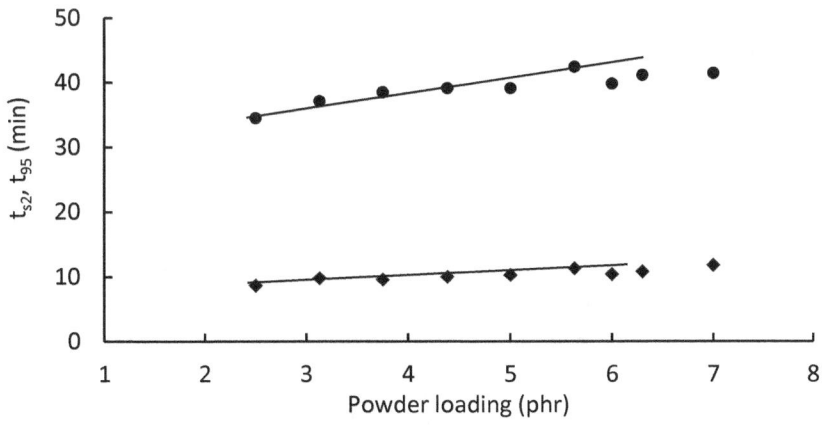

Figure 4.20 - t_{s2} (♦) and t_{95} (●) vs. powder
loading at 4 phr sulfur loading.

For the rubber compounds cured with 1 phr to 3 phr sulfur, the t_{s2} and t_{95} were averaged and plotted against the loading of sulfur (Figs. 4.21 & 4.22). The scorch time decreased by 44 % and the optimum cure time by 18 % when the loading of sulfur was raised from 1 phr to 3 phr. The addition of an increasing amount of sulfur was beneficial to these properties.

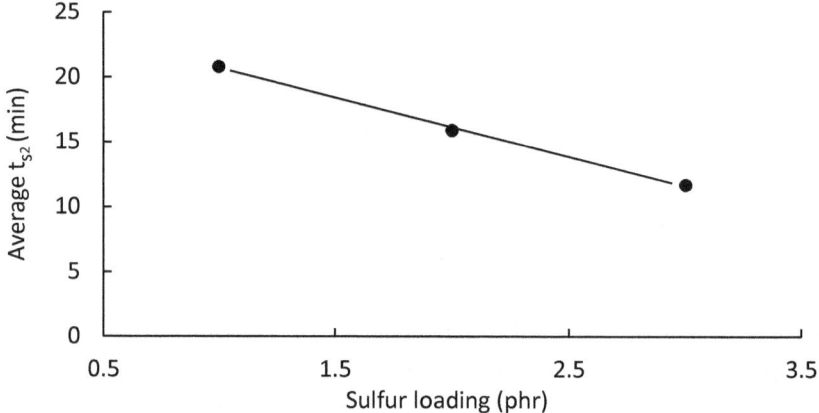

Figure 4.21 - Average t_{s2} vs. sulfur loading.

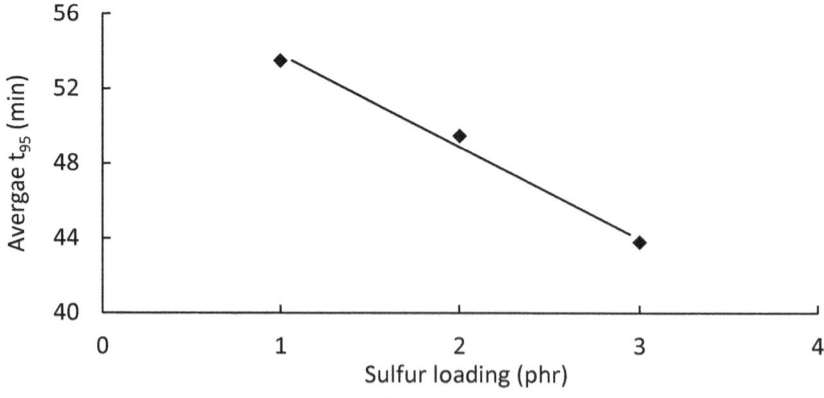

Figure 4.22 - Average t_{95} vs. sulfur loading.

The CRI vs. powder loading for the rubber compounds tested is shown in Figs. 4.23-4.26. For the rubber compounds with

1 phr to 3 phr sulfur, the cure rate, as indicated by CRI, was unchanged, but for the rubber compounds cured with 4 phr sulfur, the rate decreased by almost 13 % as the loading of the powder was increased to 7 phr.

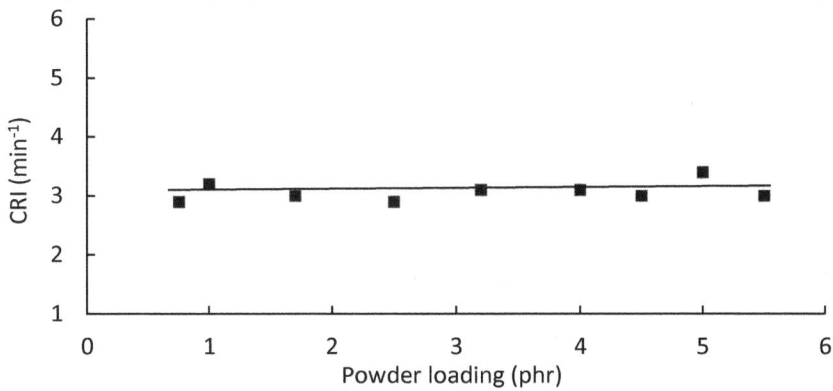

Figure 4.23 - CRI vs. powder loading at 1 phr sulfur loading.

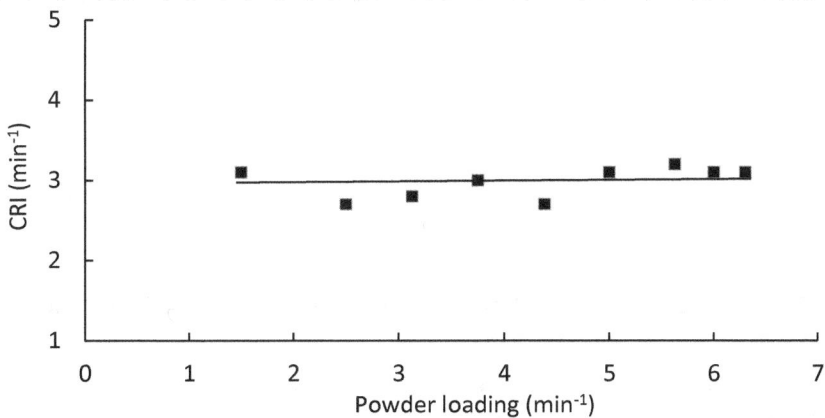

Figure 4.24 - CRI vs. powder loading at 2 phr sulfur loading.

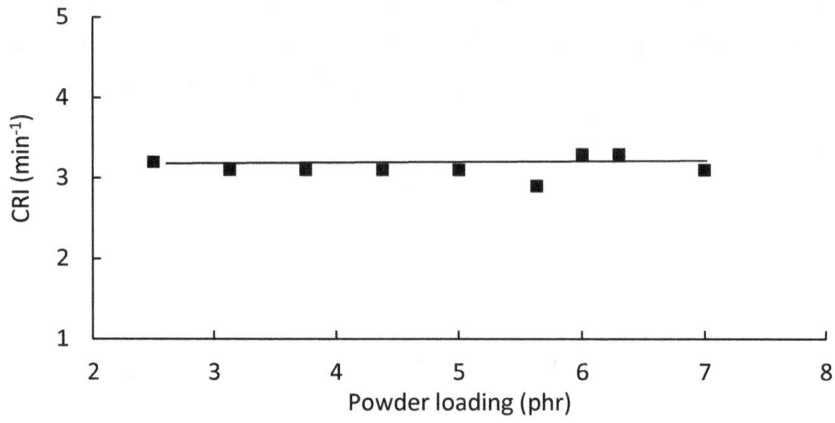

Figure 4.25 - CRI vs. powder loading at 3 phr sulfur loading.

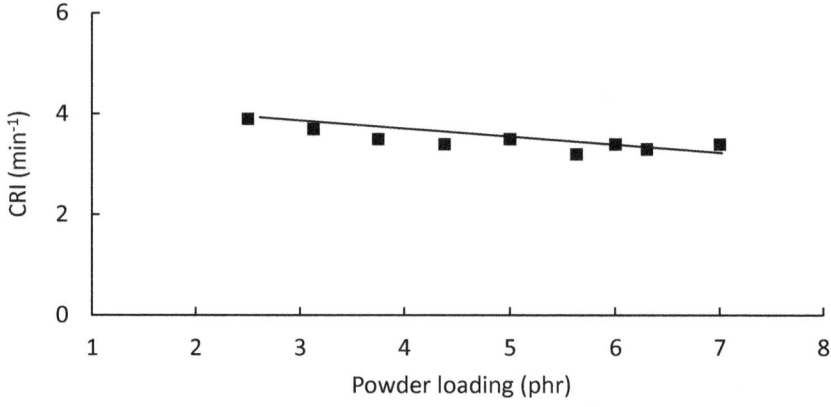

Figure 4.26 - CRI vs. powder loading at 4 phr sulfur loading.

The average CRI numbers for the rubber compounds cured with 1 phr, 2 phr, and 3 phr sulfur were plotted against the sulfur loading. It showed a constant trend, and the cure rate was not affected at all by increases in the loading of sulfur in the rubber (Fig. 4.27).

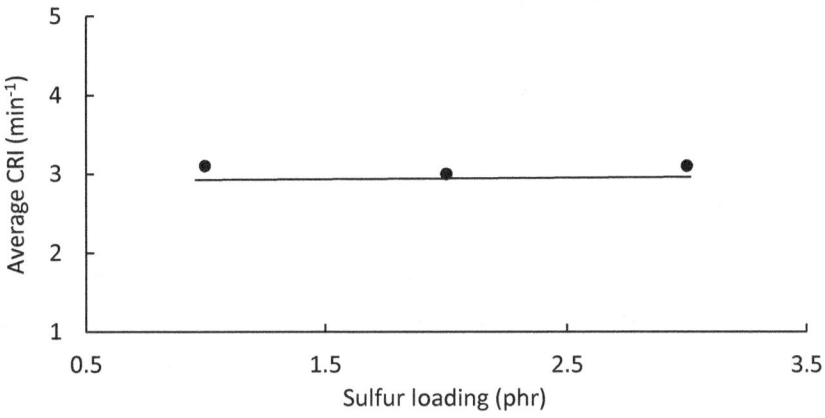

Figure 4.27 - Average CRI vs. sulfur loading.

As the results show, treating the ZnO with TBBS in an organic solvent is an efficient method for achieving an optimum cure and reducing the excessive use of these chemicals in the cure system. The most obvious benefit from the use of this powder is the ability of users to know the exact effect of the powder on the cure properties of the rubber compound containing different amounts of sulfur. Since the t_{s2}, t_{95}, and CRI were not affected by changes in the powder loading but the Δtorque was, users can change the sulfur loading to control the crosslink density for up to 3 phr sulfur without affecting the remaining cure parameters. When 1 phr, 2 phr, and 3 phr sulfur were added to the powder, the t_{s2} and t_{95} were reduced, but the CRI remained unchanged. If a high sulfur loading is preferred, i.e., 4 phr, then the cure properties will depend on the powder loading. This simplifies the selection of the powder and shortens the mixing cycle. There is also the added advantage of eliminating secondary accelerators, using a lot less ZnO, and removing stearic acid entirely from the cure system. The sulfur cure systems in industrial rubber compounds are inefficient, too costly, harmful to animal and human health and the environment, and no longer viable. A cure system based on a single TBBS/ZnO composite powder and sulfur is a more efficient way to vulcanize the rubber and reduce the excessive use of these chemicals in the cure system without compromising the mechanical properties of the rubber vulcanizate [4.3,4.4].

4.1.4 Summary

One cure system for the BR and four for the EPDM rubbers were measured using the powder with different amounts of sulfur.

For the BR, (S/powder): (4/2.5)

For the EPDM, (S/powder): (1/3.2), (2/3.75), (3/5), and (4/5.63).

4.2 References

[4.1] – P. A. Ciullo, N. Hewitt, "The rubber formulary," Noyes publications, New York, 1999, p. 181.

[4.2] – P. A. Ciullo, N. Hewitt, "The rubber formulary," Noyes publications, New York, 1999, p. 281.

[4.3] – A. Ansarifar, K. Noulta, G. W. Weaver, K. G. U. Wijayantha, "Mineral kaolin for rubber reinforcement", REF Rubber Fibres Plastics, 16(4) (2021), 182.

[4.4] – A. Ansarifar, K. Noulta, S. H. Sheikh, G. W. Weaver, K. G. U. Wijayantha, "Rubber cure system", Tire Technology International, (2021), 56.

CHAPTER 5

5 The health, safety, cost, and environmental
benefits of using the new methods for measuring
the chemical curatives for rubber vulcanization

5.1 Comparing the efficiency of the two powders in the sulfur vulcanization of NR

The two powders had different effects on the cure properties of NR at 4 phr sulfur loading. The rubber compound that was cured with the TBBS/ZnO powder had a scorch time of 4.3 min at 0.63 phr of the powder. As the loading of the powder was increased, the scorch time dipped to 3.4 min at 1.25 phr of the powder. It then rose again to 4 min when the loading of the powder reached 5.63 phr. The rubber compound that was cured with the TMTD/ZnO powder had a shorter scorch time of 2.4 min at 0.63 phr of the powder. The scorch time decreased to 2 min, when the full amount of the powder was incorporated into the rubber (Fig. 5.1).

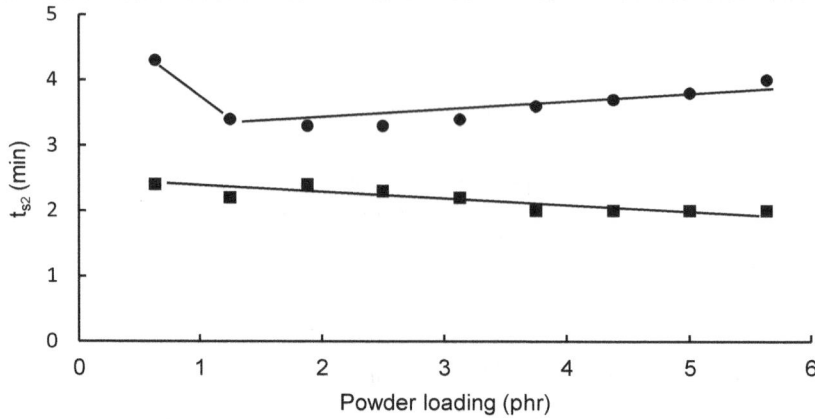

Figure 5.1 - t_{s2} vs. powder loading, (■) TMTD/ ZnO powder, (●) TBBS/ZnO powder.

The optimum cure time of the rubber compound cured with the TBBS/ZnO powder was 30.8 min at 0.63 phr of the powder. It reduced sharply to 7.4 min at 1.25 phr of the powder and remained unchanged at about 6.6-7.4 min thereafter. The rubber compound that was cured with the TMTD/ZnO powder had a much shorter optimum cure time of 3.5 min at 0.63 phr of the powder. It then increased slightly to 4 min at 5.63 phr of the powder (Fig. 5.2).

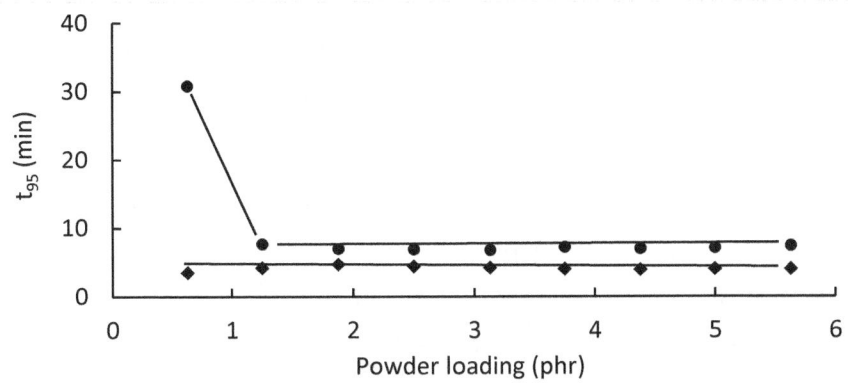

Figure 5.2 - t_{95} vs. powder loading, (♦) TMTD/ ZnO powder, (●) TBBS/ZnO powder.

For the rubber compound cured with the TBBS/ZnO powder, the cure rate, as indicated by CRI, rose sharply from 3.8 min^{-1} at 0.63 phr to 23.2 min^{-1} at 1.25 phr of the powder. It increased slowly, reaching 29.4 min^{-1} at 5.63 phr of the powder. Initially, the rubber compound that was cured with the TMTD/ZnO powder had a much faster cure, with a CRI of 91 min^{-1} at 0.63 phr of the powder. But it was reduced to about 43.5 min^{-1} to 52.6 min^{-1} when up to 5.63 phr of the powder was added (Fig. 5.3).

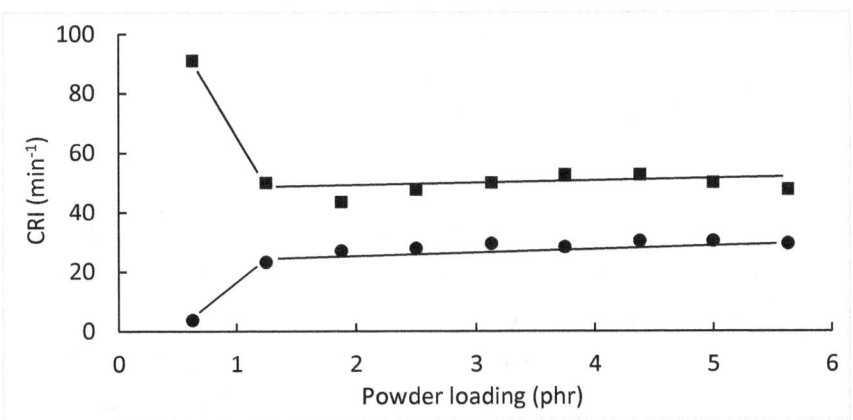

Figure 5.3 - CRI vs. powder loading, (■) TMTD/ ZnO powder, (●) TBBS/ZnO powder.

Interestingly, the powders had very similar effects on the crosslink density changes in the rubber, as shown by the Δtorque values. For the rubber compound cured with the TBBS/ZnO powder, the Δtorque rose from 22 dNm at 0.63 phr to 48 dNm at 2.5 phr of the powder. It then continued rising at a much slower rate, to about 65 dNm at 5.63 phr of the powder. Similarly, for the rubber compound cured with the TMTD/ZnO powder, Δtorque rose from 15 dNm at 0.63 phr to 49 dNm at 2.5 phr of the powder. ΔTorque continued to increase at a much slower rate, reaching about 59 dNm at 4.38 phr of the powder. Finally, it reached a slightly lower value of 57 dNm when the full amount of the powder was added to the rubber (Fig. 5.4).

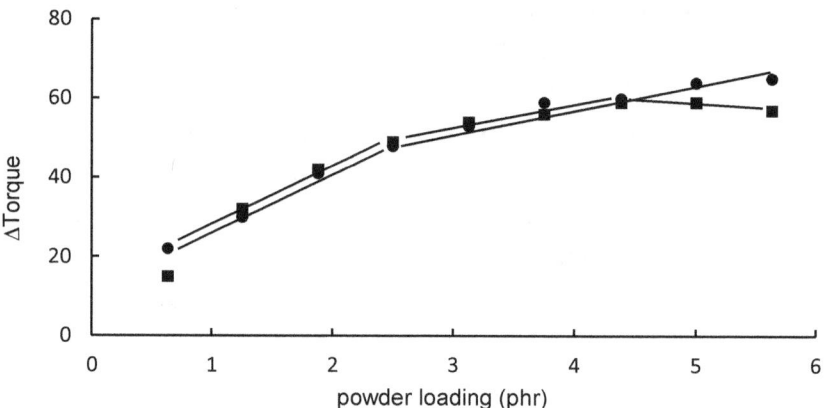

Figure 5.4 - ΔTorque vs. powder loading for the rubber compounds, (■) TMTD/ZnO powder, (●) TBBS/ZnO powder.

The optimum cure was achieved with 2.5 phr of either powder. However, the results showed that the TMTD/ZnO powder was more efficient because of the shorter scorch and optimum cure times and a much faster cure rate. The TMTD/ZnO powder in combination with some added sulfur can replace the current sulfur cure systems in industrial rubber compounds. Moreover, a significant reduction in the excessive use of TMTD and ZnO helps to minimize their adverse impact on human and animal health and the environment.

5.2 Cost, health, safety and environmental benefits

As mentioned in Chapters 2-4, the requirements for the TBBS and ZnO to fully cure the NR and EPDM rubbers differed from one method to another. Table 5.1 shows a summary of the TBBS and ZnO requirements for the NR and EPDM rubbers from methods 1 and 2.

Table 5.1: Summary of the TBBS and ZnO requirements for the NR and EPDM rubbers from methods 1 and 2.

Compound	Chemical curatives requirement for full cure		
	Method 1 (TBBS and ZnO)		
	TBBS (phr)	ZnO (phr)	Total cost in USD
NR with 4 phr sulfur	3.5	0.2	12.34
EPDM with 1 phr sulfur	1	0.075	3.565
	Method 2 (TBBS/ZnO powder)		
NR with 4 phr sulfur	0.65	1.85	6.28
EPDM with 1 phr sulfur	0.83	2.37	8.036

Market price: TBBS: USD 3.4/kg, ZnO: USD 2.2/kg

For the NR with 4 phr sulfur, the TBBS requirement was 81 wt% higher, but the ZnO requirement was 89 wt% lower, when method 1 was compared with method 2. Whereas, for the EPDM with 1 phr sulfur, the TBBS requirement was 17 wt% higher, but the ZnO requirement was approximately 3060 wt% lower, when method 1 was compared with method 2. The market price for TBBS is USD 3.3/kg to 3.4/kg, and for ZnO, USD 1.9/kg to 2.2/kg [5.1]. Taking the highest price per kg for each chemical, it was roughly 50% cheaper to cure the NR using method 2. However, for curing the EPDM, method 1 was 56% less expensive. In terms of impact on human and animal health and the environment, both TBBS and ZnO, whether they are used separately or as a single powder in the cure system, are harmful, and therefore, it is less clear which method offers a better choice. Nevertheless, it is clear from this study that the choice of the method for curing a rubber depends, to a large extent, on the chemical structure or composition of the rubber and the exact amounts of TBBS and ZnO required to cure the rubber fully. Either method is significantly more efficient than the cure systems currently in use in industrial rubber formulations.

As mentioned earlier, the sulfur cure systems in industrial rubber articles often consist of two activators and two accelerators. For instance, the cure system in an NR-based tire apex formulation has 10 phr ZnO and 2 phr stearic acid (activators), 0.6 phr TBBS and 0.25 phr tetrabenzylthiuram disulphide (TBzTD) (accelerators), and 5 phr sulfur [5.2]. The total amount of the activators and accelerators is 12.85 phr. Recall that the optimum quantity of TBBS in the powder was 350 mg/g. Thus, 26 wt% of the powder was TBBS and the remaining 74 wt% was ZnO. On this basis, 2.5 phr of the powder that was used to cure the NR compounds contained 0.65 phr TBBS and 1.85 phr ZnO. The 3.2 phr of the powder that was used to cure the EPDM compounds was made up of 0.83 phr TBBS and 2.37 phr ZnO. The cure system in the NR-based tire apex formulation could be replaced with the TBBS/ZnO powder (method 2) that uses 0.65 phr TBBS, 1.85 phr ZnO, and 4 phr sulfur. No stearic acid and no secondary accelerators were required. This will lead to a substantial decrease in the use of these chemicals and a lower sulfur requirement.

In some industrial EPDM-based rubber formulations, the cure system consists of 5 phr ZnO and 1.5 phr stearic acid (activators), 1.5 phr tetramethyl thiuram disulphide (TMTD), 0.5 phr 2-mercaptobenzothiazole (MBT) (accelerators), and 1.5 phr sulfur [5.3]. The total loading of the activators and accelerators comes to 8.5 phr. These chemicals can be replaced with 1 phr TBBS, 0.075 phr ZnO, and 1 phr sulfur (method 1). Both methods offer a baseline and a starting point from which a suitable cure system can be selected and then tailored for optimum efficiency.

Exposure to ZnO can cause a flu-like illness with symptoms of metallic taste in the mouth, headache, fever, chills, aches, chest tightness, and cough. This is highly damaging to the developing fetus and harmful to aquatic life [5.4]. There are also serious environmental concerns related to the use of discarded old tires. Zinc leaching is a major concern when tire crumb rubber is used in certain environmental settings. For example, in civil engineering applications, it was found that zinc leaching from tire crumb rubber increased with longer exposure time. This caused significant environmental pollution, and when the current volume of tire crumb rubber usage is considered, the damage to the environment becomes clear [5.5]. The accelerator TBBS may cause an allergic skin reaction and is very toxic to aquatic life with long-lasting effects [5.6]. There are also serious safety issues with the TMTD accelerator. TMTD is classified as being toxic to aquatic organisms and is irritating to the eyes, respiratory system, and skin [5.7]. Moreover, TMTD contains secondary amine that produces nitrosamines during mixing, vulcanization, and storage of rubber articles [5.8]. Nitrosamines are listed as carcinogenic according to the "Technical Rules for Dangerous Substances" (TRGS 522) [5.9]. When either powder is used to cure the NR, BR, and EPDM rubbers, it will significantly reduce the use of TBBS and ZnO in vulcanization. This will help to minimize the risk to health and improve safety at work. A major reduction in the use of TBBS and ZnO in the cure system will minimize the air pollution and soil contamination caused by the leaching of the rubber articles into the environment at the end of their useful service life.

5.3 Compound integrity and structural homogeneity of rubber compounds

Compound integrity is a major concern for rubber compounders. The mixing of chemicals with raw rubber is meant to ensure full distribution and dispersion in the rubber to achieve homogeneous crosslinking and crosslink distribution, but often this is not the case. Cure efficiency is only possible when the chemicals react together at high temperatures, and the distribution and dispersion of the curatives in rubber is a random process at the best of times. Statistically, the probability of all the chemicals coming together at the right weight ratios everywhere in the rubber to react and produce crosslinks uniformly throughout the rubber matrix is very low and varies significantly from one rubber compound to another. The more chemicals are mixed with rubber, the more problematic the dispersion becomes and the less homogeneous the crosslink distribution in the vulcanized rubber will be. This problem has been known for many years, and work has been done to remedy it. For example, to disperse solid fillers such as synthetic precipitated silica more efficiently in rubber, a highly dispersive grade of the filler with a high surface area has been developed [5.10]. Progress has been limited in the case of chemical curatives such as ZnO and TBBS. Long mixing times can improve the quality of solid dispersion in rubber, but this process is energy-intensive and expensive. Moreover, some chemicals have melting points below the curing temperature of most rubber compounds, which is typically somewhere between 140 to 220°C [5.11]. For example, the melting temperature of TBBS is 105°C. At this temperature, the TBBS melts inside the rubber during curing and then solidifies, forming large aggregates when the vulcanizate is cooled down in storage at room temperature (Fig. 5.5).

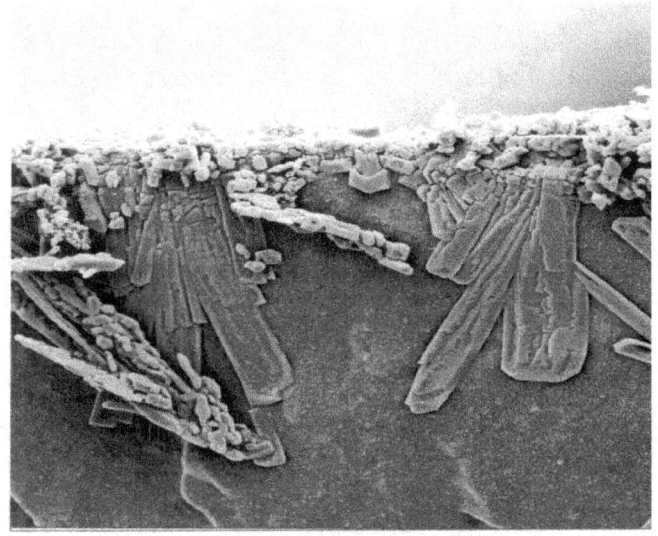

10 µm

Figure 5.5 – Scanning electron micrograph (SEM) (cross-section) from the freeze-fracture tests showing aggregated TBBS inside the rubber and migrated aggregates on the rubber surface.

In storage, the aggregates migrate through the rubber (Fig. 5.6) and reach the surface, forming blooms (Fig. 5.7).

bloom

cracks

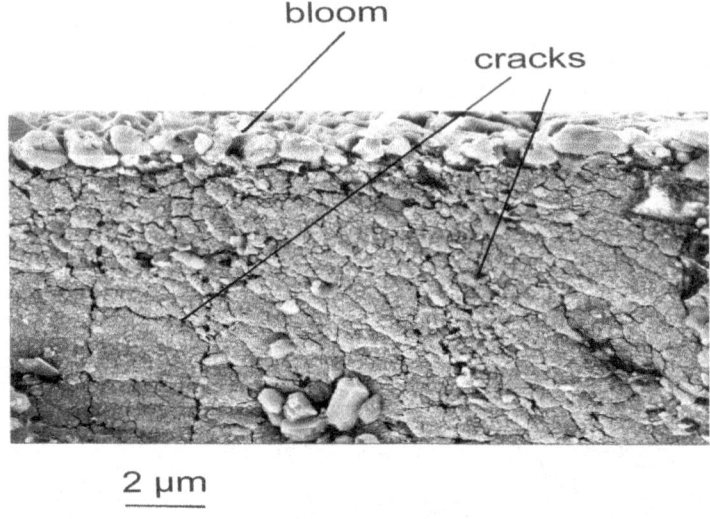

2 µm

Figure 5.6 – SEM (cross-section) from the freeze-fracture tests showing TBBS bloom on the rubber surface and cracks below the bloomed layer in the rubber.

100 µm

Figure 5.7 – SEM showing needle-shaped objects (TBBS) on the surface of the unstrained rubber.

The migration of the aggregates through the rubber matrix is highly damaging to the internal structure of the rubber and leaves cracks and large voids behind that weaken the rubber (Fig. 5.8). Likewise, the unreacted ZnO diffuses through the rubber matrix and reaches the surface, causing contamination [5.5].

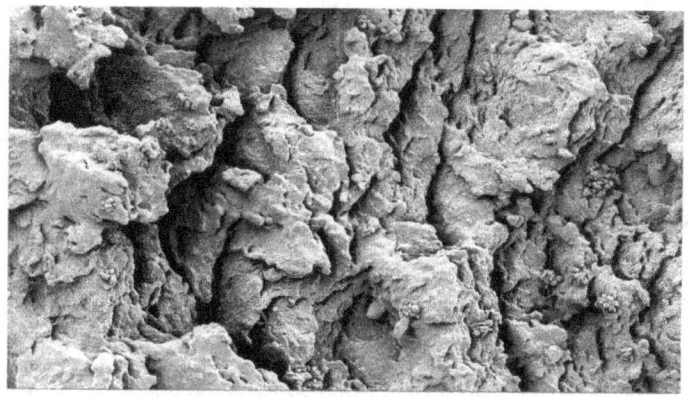

40 μm

Figure 5.8 – SEM photograph showing a fracture surface after a cyclic fatigue test. Note the cracks and extensive damage to the rubber due to the migration of TBBS aggregates through the rubber matrix to the surface.

The migration of the unreacted chemicals to the surface and blooming can be prevented when fewer chemicals with high melting points are used. Method 1 uses TBBS, ZnO, and sulfur to cure the rubber. No stearic acid and secondary accelerators were used. Method 2 uses either TBBS/ZnO powder and sulfur, or TMTD/ZnO powder with or without added sulfur to cure the rubber. Although the former uses fewer chemicals than the current cure systems in industrial rubber compounds, the dispersion of the TBBS, ZnO, and sulfur in the rubber matrix still poses a challenge. But the latter uses only one powder instead of the more traditional two accelerators, two activators, and sulfur to cure the rubber. As mentioned in section 3.2.4, the amount of the TMTD/ZnO powder needed to fully cure the rubber was at a 12 phr level. This was excessive. There are health and safety concerns related to the use of TMTD in vulcanization. Recent studies have shown that TMTD can be replaced with a safer sulfur-donor accelerator [5.8], though the efficiency of the new accelerator in releasing sulfur to cure the rubber must be evaluated first before it can be used to produce a suitable powder like TMTD/ZnO for use in industrial rubber compounds. TMTD is a highly efficient sulfur-donor accelerator [5.12]. If a single powder like TMTD/ZnO is ever to be used in industrial rubber compounds, it must release sulfur efficiently during curing and be safe for use on a regular basis. It may well be within the reach of rubber chemists to develop such a product soon and transform the way rubber articles are vulcanized. The benefits for the users of rubber chemicals and the rubber industry will be immense.

5.4 A software program for selecting cure systems for NR

A software program has been developed to select cure systems for the NR. The rubber was mixed with different amounts of the TBBS/ZnO powder and sulfur to produce rubber compounds. The rubber compounds were tested in a curemeter, and the results were used to produce a database. The database was incorporated into the program. The program performed some tasks to produce useful information for selecting a range of cure systems for the rubber.

5.4.1 The database - Effects of an increasing loading of TBBS/ZnO powder on the crosslink density changes in the rubber with different amounts of sulfur

To run the software program, the cure properties of the rubber compounds with different amounts of the powder and sulfur were measured to create a database. The loading of sulfur in the rubber compounds was increased from 1 phr to 5.5 phr and that of the powder from 0.50 phr to 10.4 phr to determine their effects on the cure properties (64 rubber compounds were made). The temperature of the rubber compounds during mixing was 52°C-62°C. The t_{s2}, t_{95}, CRI, M_H, and M_L of the rubber compounds were measured. ΔTorque was plotted against the loading of the powder to measure the optimum amount of the powder required for vulcanizing the rubber compounds containing different quantities of sulfur. Six rubber compounds with the optimum powder loading were selected for further work (Table 5.2). The Δtorque vs. powder loading figures from which the data in Table 5.2 were taken will be discussed later.

Table 5.2: Formulations and cure properties of the rubber compounds at the optimum powder loading. Data taken from Figs. 5.9a-5.9b.

Formulation (phr)	Compound no					
NR	100	100	100	100	100	100
Sulfur	1	2	3	4	5	5.5
Optimum powder loading	1	1.4	1.88	2.5	3.13	5
TBBS/ZnO	0.26/0.74	0.36/1.04	0.49/1.39	0.65/1.85	0.81/2.32	1.3/3.7
	Curemeter test results at 160°C					
M_L (dNm)	16	15	15	15	14	14
M_H (dNm)	33	47	58	63	74	95
ΔTorque (dNm)	17	32	43	48	60	81
t_{s2} (min)	7.2	5.1	4.2	3.3	3.8	4.2
t_{95} (min)	12.2	9.2	7.9	6.9	7.2	7.7
CRI (min^{-1})	20	24.4	27.0	27.8	29.4	28.6

TBBS/ZnO: 26wt%/74wt%

The cure traces of the rubber compounds in Table 5.2 are presented in Fig. 5.9.

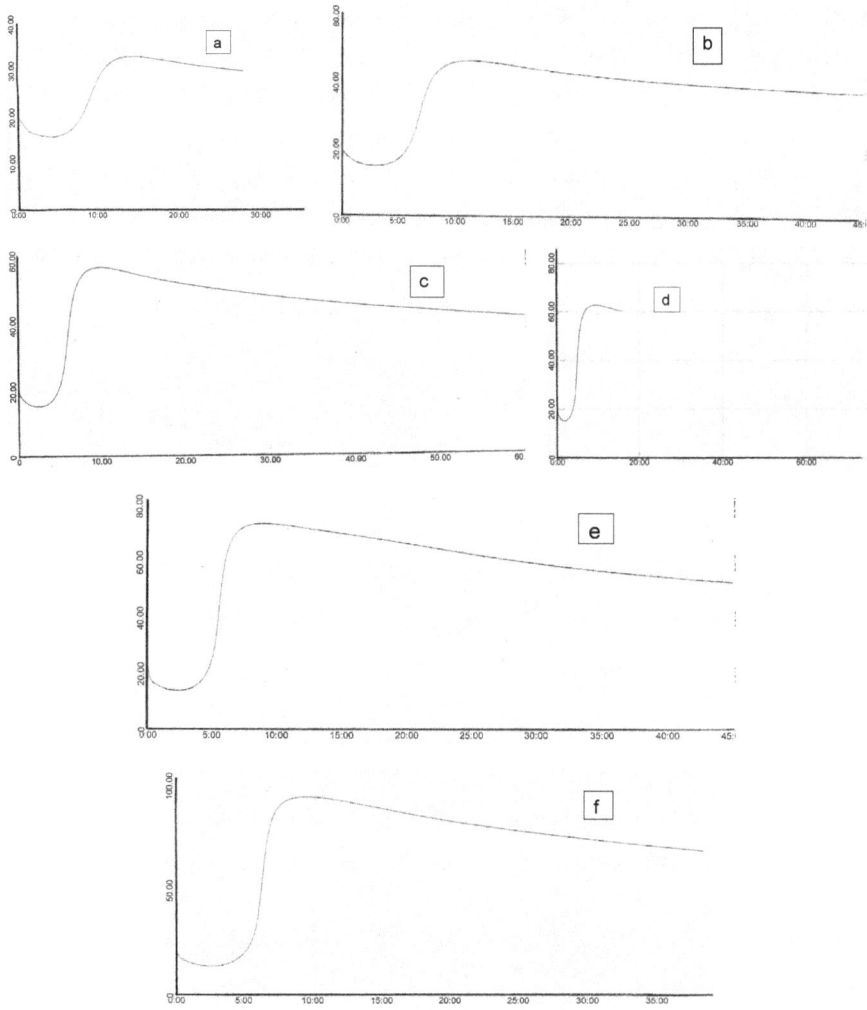

Figure 5.9 - Cure traces (Torque (dNm) vs. time (min)) of the rubber compounds with an increasing loading of sulfur. a) compound with 1 phr sulfur and 1 phr powder, b) compound with 2 phr sulfur and 1.4 phr powder, c) compound with 3 phr sulfur and 1.88 phr powder, d) compound with 4 phr sulfur and 2.5 phr powder, e) compound with 5 phr sulfur and 3.13 phr powder, f) compound with 5.5 phr sulfur and 5 phr powder.

Figures 5.10 to 5.15 show Δtorque as a function of the powder loading for the rubber compounds tested. For the rubber compound with 1 phr sulfur (Fig. 5.10), Δtorque increased from 8 dNm to 17 dNm when the powder loading was raised from 0.5 phr to 1 phr. ΔTorque continued rising at a much slower rate, reaching about 33 dNm at 3 phr of the powder. It then reached 42 dNm at 5.4 phr of the powder. The 1 phr of the powder was enough to cause the sulfur to react with the rubber to form stable covalent crosslinks between the rubber chains.

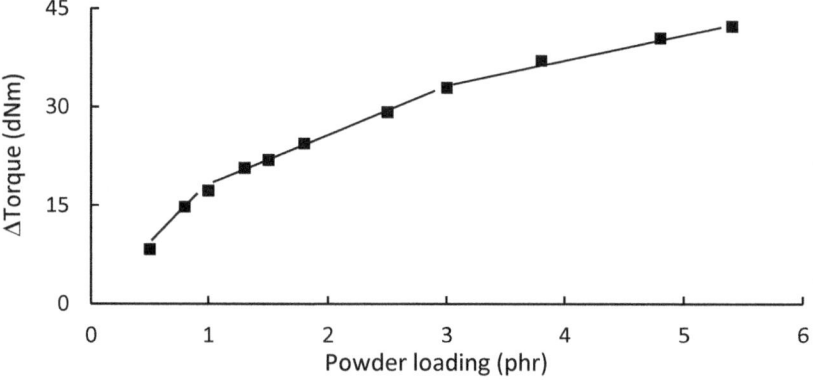

Figure 5.10 - ΔTorque vs. TBBS/ZnO powder loading for the rubber compound with 1 phr sulfur.

For the rubber with 2 phr sulfur (Fig. 5.11), Δtorque increased sharply from 13 dNm to 32 dNm when the loading of the powder was raised to 1.4 phr. ΔTorque then attained a value of 53 dNm when the full amount of the powder, i.e., 5.63 phr, was added to the rubber.

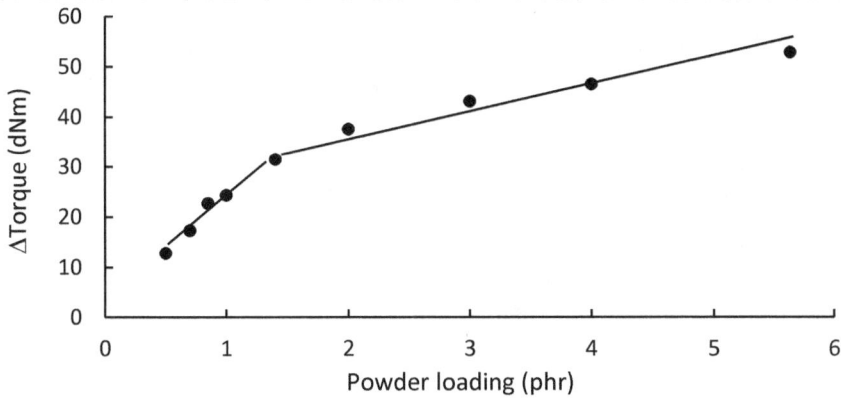

Figure 5.11 – ΔTorque vs. TBBS/ZnO powder loading for the rubber compound with 2 phr sulfur.

A similar trend was observed for the rubber with 3 phr sulfur (Fig. 5.12). For this rubber, Δtorque rose from 20 dNm to 42 dNm with 1.88 phr of the powder. It then continued to increase to about 63 dNm when the full amount of the powder was incorporated into the rubber.

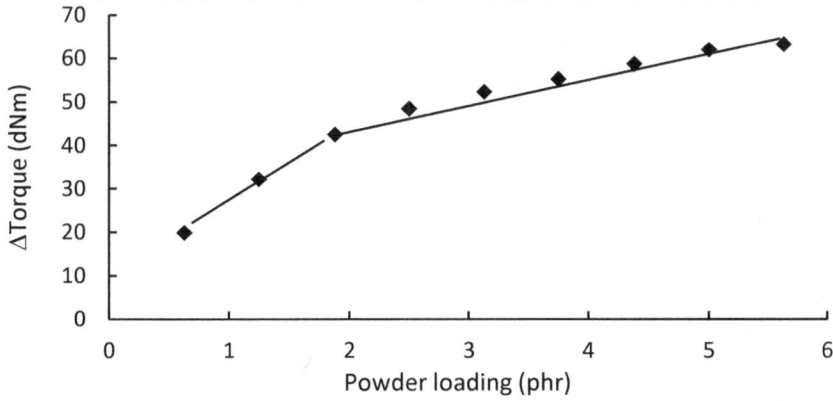

Figure 5.12 – ΔTorque vs. TBBS/ZnO powder loading for the rubber compound with 3 phr sulfur.

As expected, Δtorque showed improvement when the loading of sulfur was raised to 4 phr (Fig. 5.13). For this rubber, Δtorque increased from 22 dNm to 48 dNm when the loading of the powder reached 2.5 phr. It then continued rising to 65 dNm when the loading of the powder was raised to 5.63 phr.

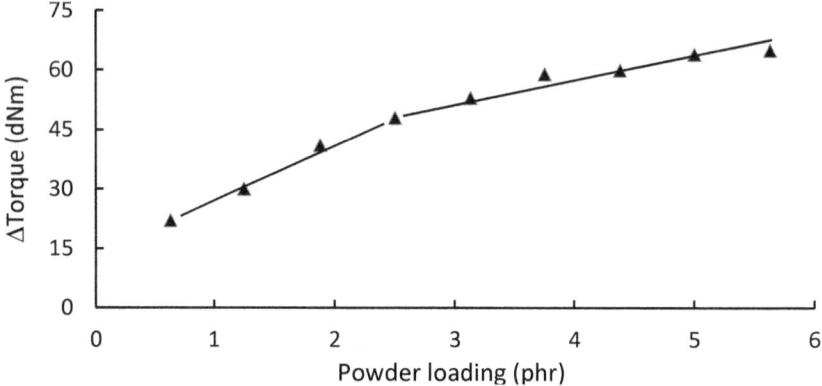

Figure 5.13 – ΔTorque vs. TBBS/ZnO powder loading for the rubber compound with 4 phr sulfur.

Similarly, for the rubber with 5 phr sulfur (Fig. 5.14), Δtorque rose from 27 dNm at 0.63 phr to 60 dNm at 3.13 phr of the powder. ΔTorque continued rising at a slower rate to about 83 dNm when the loading of the powder reached 5.63 phr.

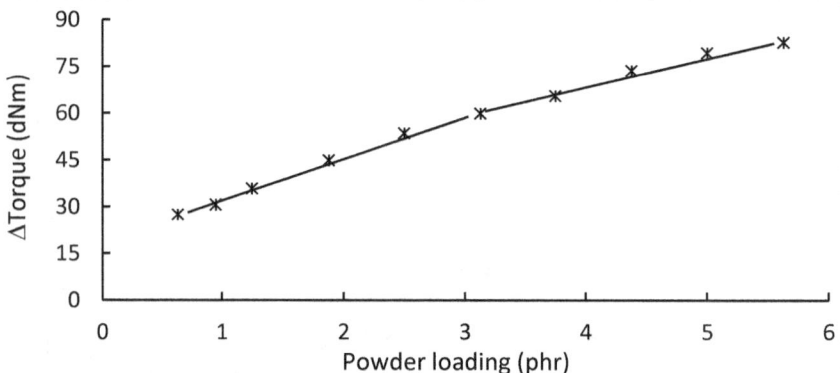

Figure 5.14 – ΔTorque vs. TBBS/ZnO powder loading for the rubber compound with 5 phr sulfur.

Finally, the Δtorque for the rubber with 5.5 phr sulfur (Fig. 5.15) showed a large improvement from 31 dNm to 81 dNm as the loading of the powder was raised from 0.63 phr to 5 phr. But the rate of increase slowed down, reaching about 112 dNm when the loading of the powder was raised to 10.4 phr.

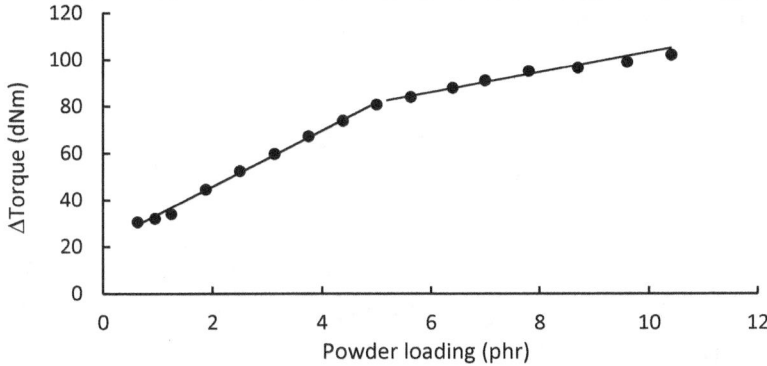

Figure 5.15 – ΔTorque vs. TBBS/ZnO powder loading
for the rubber compound with 5.5 phr sulfur.

For curing the rubber, the optimum loading of the powder increased linearly as a function of the sulfur loading for up to 5 phr. It then rose sharply when an additional 0.5 phr of sulfur was added (Fig. 5.16).

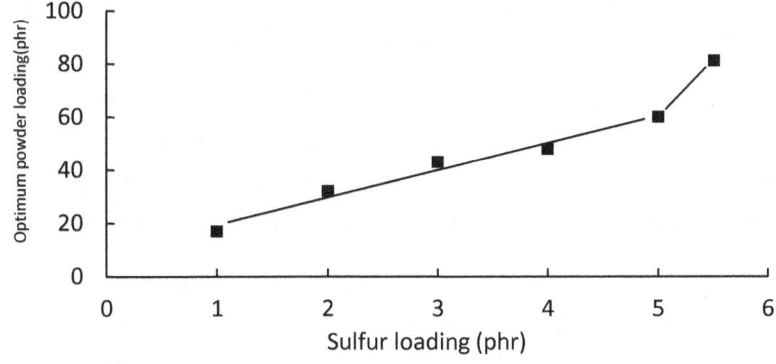

Figure 5.16 - Optimum TBBS/ZnO powder
loading vs. sulfur loading.

In the NR-based industrial rubber formulations, the loading of sulfur does not exceed 3 phr [5.2,5.11]. Therefore, the correlation between the optimum loading of the powder and the sulfur loading in Fig. 5.16 is valid for industrial rubber formulations that use TBBS and ZnO in the cure system. After these measurements were completed, the results (Table 5.2) were used to create a database. The database was incorporated into the software program.

5.4.2 Development of a software program for cure systems for NR

The following steps were taken in chronological order to store and process the cure test results from Table 5.2 in Section 5.4.1. and run the program.

- The data from Table 5.2 was copied to a digital spreadsheet.

- The relationships between the sulfur loading, powder loading, t_{s2}, t_{95}, CRI, and Δtorque were examined and then modelled using a developed algorithm. The developed algorithm processed the sulfur loading and outputted the values of the t_{s2}, t_{95}, CRI, M_H, M_L, and Δtorque properties, and then created a function that took any numeric input from the sulfur trend line and produced the subsequent numerical value on the existing trend line.

- The method was then replicated for each of the properties provided in Table 5.2. Individual algorithms were developed for each property to convert numerical sulfur loading inputs into different outputs, i.e., one algorithm for each property.

- A webpage was created using the information in the spreadsheet from Table 5.2, and the algorithms for each property were written in JavaScript. The digital data spreadsheet was written into the webpage, displaying information in a table for the user (Fig. 5.17a).

- A program for a Radar chart was subsequently created and produced on the webpage with spokes for each of the following properties: t_{s2}, t_{95}, CRI, M_L, M_H, Δtorque, powder loading, and sulfur loading (Fig 5.17b).

- Both a numerical input text box and a slider were used to input continuous data for sulfur loading between reasonable values, for example, between 1 phr and 2 phr.

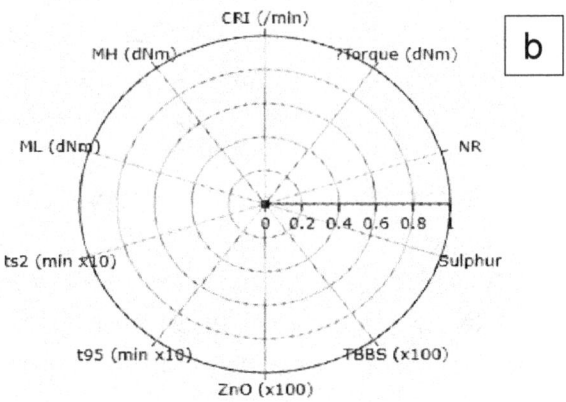

Figure 5.17 – Typical webpage, displaying information for user (a), and typical Radar chart created on the webpage with spokes for each of the properties (b).

In summary, the program was written in HTML. Initial prototypes of the program were developed on Excel. Using the data from Table 5.2, a function was created that converted the discrete data points into continuous extrapolated data points between the points and provided a resolution of 1 dp. Functions were created for each set of the provided points. For example, one function for extrapolating between 1 phr and 2 phr of the sulfur loading, another function for extrapolating between 2 phr and 3 phr of the sulfur loading. This was repeated for the remaining sulfur loadings and was performed individually for all the variables. These functions were used to create algorithms that received sulfur loading inputs in phr to produce discrete extrapolated outputs for all the variables. These algorithms were written in JavaScript, and the outputs of these functions were displayed on a generated Output Box and Radar Chart.

5.4.3 Use of the software program to select cure systems for the NR

As mentioned earlier, the digital data spreadsheet was written into the webpage, displaying information in a table for the user (Fig. 5.17a) and a program for a Radar chart was subsequently created and produced on the webpage with spokes for each of the properties (Fig. 5.17b). Each algorithm processed the inputted sulfur loading for each reading and outputted a value for each property. For instance, a sulfur loading of 1.5 phr, outputted values of TBBS: 0.31; ZnO: 0.89; Δtorque: 24.5; t_{s2}: 6.7; t_{95}:10.7, and CRI: 22.2. These values were subsequently displayed in a table after the program had been executed (Fig. 5.18a) and on a Radar chart for a more visual interface (Fig. 5.18b).

Formulation(phr)	
CRI	22.2
DeltaTorque	24.5
NR	100
Sulphur	1.5
TBBS	31
ZnO	89
t95	107
ts2	67

a

Final Cure System at a Touch of a Button

Demonstration of Array and semi-working Radar Chart

[refresh] [draw]

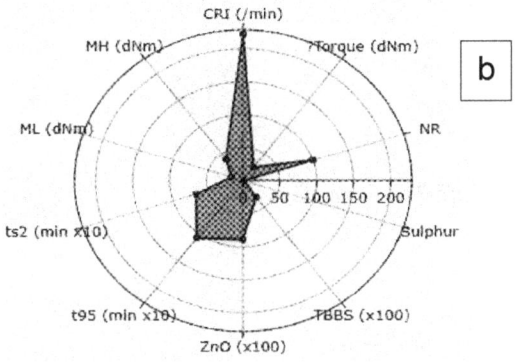

Figure 5.18 – Webpage, displaying information for the rubber at 1.5 phr sulfur loading after the program was executed (a), and Radar chart created with spokes for each of the properties (b).

For a sulfur loading of 4.5 phr, outputted values of TBBS:0.73; ZnO:2.085; Δtorque: 54; t_{s2}:3.55; t_{95}:7.05, and CRI:28.6, were produced. The results were subsequently displayed in a table (Fig. 5.19a) and on a Radar chart (Fig. 5.19b).

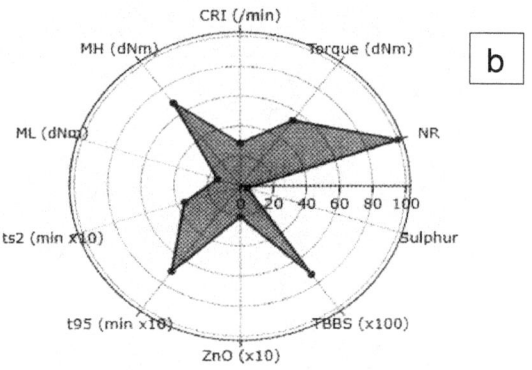

Figure 5.19 – Webpage, displaying information
for the rubber at 4.5 phr sulfur loading after the
program was executed (a) and Radar chart created
with spokes for each of the properties (b).

The numbers for the TBBS and ZnO loadings in the table (Fig. 5.18a) were divided by a factor of 100, and for those of t_{s2} and t_{95}, by a factor of 10, when the Radar chart was constructed (Fig. 5.18b). For the sake of clarity of presentation, the original TBBS and ZnO, and t_{s2} and t_{95} values from Table 5.2, were multiplied by factors of 100 and 10, respectively. For the 4.5 phr sulfur loading, the TBBS and ZnO numbers from the table (Fig.

5.19a) were divided by factors of 100 and 10, respectively, and those of t_{s2} and t_{95}, by a factor of 10, as indicated in the Radar chart (Fig. 5.19b). As Figs. 5.18 and 5.19 show, the software program helped to select cure systems for the rubber. This was achieved by inserting a value for the sulfur loading into the program and then performing the tasks described above. For example, when the information in the table (Fig. 5.18a) was judged to be unsuitable, e.g., t_{95} was too long or too short, a new value for the sulfur loading was inserted into the program and the tasks were repeated. This can be repeated several times until a sulfur loading is found at which t_{95} may be desirable for the user. Once this software program is fully implemented into the compounding process, a major reduction in the processing time may be achieved.

To summarise, the NR containing different amounts of sulfur was cured with a single powder. The optimum loadings of the powder required to cure the rubber compounds with 1 phr, 2 phr, 3 phr, 4 phr, 5 phr, and 5.5 phr sulfur were determined. Using a JavaScript program, the cure test results were stored and processed to develop a software program to help with the selection of cure systems for the rubber. The software program offered significant advantages over the traditional methods of selecting cure systems for the rubber. It provided useful information on the t_{s2}, t_{95}, CRI, M_H, M_L, and Δtorque and the TBBS and ZnO requirements at 1 phr to 5.5 phr sulfur loadings. By doing so, it removed the need to mix and test the raw rubber with the TBBS and ZnO repeatedly to find a rubber compound with the ideal cure properties. Furthermore, using the powder reduced the excessive use of TBBS and ZnO in the vulcanization of the rubber. It offered a significant reduction in cost, a major improvement in health and safety at work, and minimized the damaging impact of these chemicals on the environment. All the indications are that a software program like the one developed here can select and optimise cure systems for the rubber at a desired sulfur loading very quickly. This can potentially reduce the time in the compounding stage of the rubber processing.

5.4.4 Summary

- The NR was cured with 1 phr sulfur and 1 phr powder, 2 phr sulfur and 1.4 phr powder, 3 phr sulfur and 1.88 phr powder, 4 phr sulfur and 2.5 phr powder, 5 phr sulfur and 3.13 phr powder, and 5.5 phr sulfur and 5 phr powder. The results were used to develop a database for a software program to help with the selection of cure systems for the NR. This procedure can be repeated for other industrially important rubbers.

- The correlation between the optimum powder loading and the sulfur loading was linear for up to 5 phr sulfur. This is a useful information for measuring the optimum powder requirement for curing the rubber at any loading of sulfur at or below 5 phr.

5.5 References

[5.1] – http://www.alibaba.com. Date visited: 15.08.2020.

[5.2] – R. N. Datta, "Rubber curing systems," Rapra Report 144, 12/12 (2002), 37.

[5.3] – R. N. Datta, "Rubber curing systems," Rapra Report 144, 12/12 (2002), 16.

[5.4] – https://nj.gov/health/eoh/rtkweb/documents/fs/2037.pdf Date visited: 17-11-2021

[5.5] – https://sustainablewayland.org/wp- content/ uploads/2021/03/ZincFromCrumbRubberMays.pdf Date visited: 17-11-2021

[5.6] – https://americasinternational.com/wp-content/ uploads/2018/08/TBBS-GRANULES-SDS-A.I..pdf Date visited 17-11-201

[5.7] – http://sunboss.ca/files/products/tmtd-579140.pdf Date visited 18-11-21

[5.8] – Sheth, R. N. Desai, "Replacing TMTD with nitrosamine free TBzTD accelerator in curing rubber". Inter. J. Sci. Res. Dev., 1 (3), 2013, p. 532-535.

[5.9] – https://www.fda.gov/media/147331/download Date visited 18-11-21

[5.10] – https://patents.google.com/patent/US20060137575A1/en Date visited: 19-11-2021

[5.11] – P. A. Ciullo, N. Hewitt, "The rubber formulary", Noyes Publications, New York 1999.

[5.12] – F. Saeed, A. Ansarifar, R. J. Ellis, Y. Haile-Meskel, M. Shafiq Irfan, "Two advanced styrene-butadiene/polybutadiene rubber blends filled with a silanized silica nanofiller for potential use in passenger car tire tread compound", J. Appl. Polym. Sci., 123, (2012), p. 1518-1529.

CHAPTER 6

6.1 Summary
6.1.1 Method 1
6.1.2 Method 2
6.2 A future strategy for using chemical curatives efficiently in vulcanization
6.3 Outcome

6.1 Summary

The two single powders vulcanized the NR, BR, and EPDM rubbers with and without added sulfur very efficiently. The powders can potentially replace the current methods for sulfur vulcanization.

6.1.1. Method 1

Method 1 measured the exact requirements for the TBBS and ZnO to vulcanize the NR, BR, and EPDM rubbers at different sulfur loadings (Table 6.1).

Table 6.1 – Summary of the cure systems for the
NR, BR and EPDM rubbers with method 1.

Cure system	NR			
(S/ TBBS/ZnO) (phr/phr/phr)	(1/1.5/0.2)	(2/1.5/0.3)	(3/1.5/0.25)	(4/3.5/0.2)
	BR			
(S/ TBBS/ZnO) (phr/phr/phr)	(0.5/1.75/0.2)		(1/3/0.2)	
	EPDM			
(S/ TBBS/ZnO) (phr/phr/phr)	(1/1/0.075)			

No stearic acid and secondary accelerators were needed to complete the cure.

6.1.2. Method 2

Method 2 measured the optimum amounts of the TBBS/ZnO and TMTD/ZnO powders for vulcanizing the NR and EPDM rubbers at different sulfur loadings (Table 6.2). The TBBS/ZnO powder was made of 26 wt% TBBS and 74 wt% ZnO. The TMTD/ZnO powder was made of 28.6 wt% TMTD and 71.4 wt% ZnO. No stearic acid and secondary accelerators were needed to complete the cure.

The measurements in Tables 6.1 and 6.2 can be used as a base-line to develop and tailor new cure systems for the NR-based, BR-based and EPDM-based industrial rubber articles.

Table 6.2 - Summary of the cure systems for the
NR, BR and EPDM rubbers with method 2.

Cure system	NR					
(S/powder (TBBS/ZnO)) (phr/phr)	(1/1)	(2/1.4)	(3/1.88)	(4/2.5)	(5/3.13)	(5.5/5)
	EPDM					
(S/powder (TBBS/ZnO)) (phr/phr)	(1/3.2)		(2/3.75)		(3/5)	(4/5.63)
	NR					
(TMTD/ZnO powder) (phr)	(12)					
(S/powder (TMTD/ZnO)) (phr/phr)	(4/2.5)					

The linear correlation between the optimum TBBS/ZnO powder
loading and the sulfur loading can be used to calculate the
exact amount of the powder needed for curing the NR-based
and EPDM-based industrial rubber articles for up to 5 phr sulfur
(Fig. 6.1). This will greatly simplify the measurements of these
additives for vulcanization.

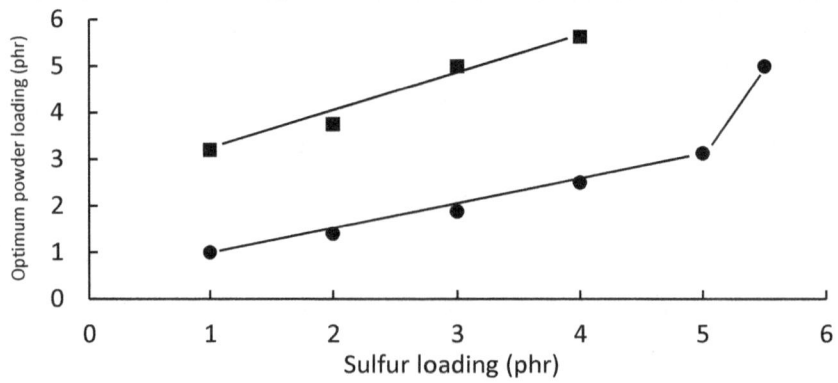

Figure 6.1 – Optimum powder loading vs.
sulfur loading. (■) EPDM, (●) NR.

126

6.2 A future strategy for using chemical curatives efficiently in vulcanization

A future strategy for compounding TMTD and ZnO with raw rubber must consider using a single additive like the TMTD/ZnO powder. A safer sulfur-donor accelerator will be required to replace the TMTD accelerator that is classified as carcinogenic at certain concentrations. The TMTD accelerator has 13 wt% sulfur available to react with rubber and a melting point of 156°C. The new accelerator (X) must be less harmful, environmentally friendly, and ideally contain more than 13 wt% sulfur to react with rubber. Furthermore, it should have a higher melting point to avoid melting and forming aggregates in the rubber during high-temperature curing. The aggregates migrate to the rubber surface, causing damage to the internal structure of the rubber, forming bloom and surface contamination. The availability of an X/ZnO powder will open a new chapter in the sulfur vulcanization of industrial rubber articles and make the rubber industry significantly more competitive and a lot safer and environmentally more appealing. The development of such a product will be challenging for rubber chemists, but past successes have shown that such products can emerge in the market soon. So, it may not be too long before a suitable X/ZnO product is available to the rubber industry. In addition to the two powders, a software program manages the selection of the cure systems effectively and quickly to improve the compounding stage of these chemicals with the raw rubber. The software program requires an input for sulfur, and then it provides the cure parameters such as t_{s2}, t_{95}, CRI, M_H, M_L, Δtorque, and the powder requirement. This is important for compounding the rubber and will reduce the mixing time and make the process very cost-effective. In addition, the use of a single powder and the software program eliminates the need to mix and test the raw rubber with TBBS and ZnO repeatedly to find a compound with ideal cure properties. It provides a rapid way of selecting the right cure systems for a rubber. The next stage is to implement the two methods and commercialise the software program for use by the rubber industry.

6.3 Outcome

- There are two highly efficient methods available for reducing the excessive use of the TBBS, TMTD, and ZnO chemicals in sulfur vulcanisation. In some cases, the chemicals in the cure system are reduced by as much as 80 wt% whilst the cure remains very effective.

- A software program that uses the cure properties of a rubber compound as a database has been developed. The program selects and optimises cure systems for NR at a selected loading of sulfur. This can potentially reduce the time in the compounding stage of the rubber processing.

It is no longer viable to use two accelerators, two activators, and sulfur to vulcanize rubber. Combining the TBBS and ZnO additives into a single powder is a highly efficient method of using them.

7 APPENDIX

7.1 List of References

[1.1] – L. Bateman, "The chemistry and physics of rubber-like substances", Studies of the natural rubber producers' research association, Maclaren & Sons Ltd. London, John Wiley & Sons. New York. 1963, chapter 1.

[1.2] – C. Hepburn, R. J. W. Reynolds, "Elastomers: Criteria for engineering design", Applied Science Publishers Ltd, London, 1979, Chapters 1&12. (ISBN 0-85334-809-X)

[1.3] – https://www.deltarubber.co.uk/rubber-industry Date visited 30-11-2021

[1.4] – S. C. Nyburg, "X-ray determination of crystallinity in deformed natural rubber", Brit J. Appl. Phys., 1954, 5, 321.

[1.5] – L. Bateman, "The chemistry and physics of rubber-like substances", Studies of the natural rubber producers' research association, Maclaren & Sons Ltd. London, John Wiley & Sons. New York. 1963, chapter 9.

[1.6] – P. A. Ciullo, N. Hewitt, "The rubber formulary", Noyes Publications, NY 1999.

[1.7] – W. Hofmann, "Rubber technology handbook," Hanser Publishers, New York, 1989, Chapter 3. (ISBN 3-446-14895-7) (Hanser) Pp.

[1.8] – A. D. Roberts, "Natural rubber science and technology", Oxford science publications, 1988, Chapter 19 (ISBN 0-19-855225-4).

[1.9] – W. Hofmann, "Vulcanization and vulcanizing agents", Maclaren and Sons Ltd, London, 1967, Chapter 2.

[1.10] – W. Hofmann, "Vulcanization and vulcanizing agents", Maclaren and Sons Ltd, London, 1967, Chapter 3.

[1.11] – W. Hofmann, "Rubber technology handbook," Hanser Publishers, New York, 1989, Chapter 4. (ISBN 3-446-14895-7) (Hanser) Pp.

[1.12] – W. Hofmann, "Vulcanization and vulcanizing agents", Maclaren and Sons Ltd, London, 1967, Chapter 4.

[1.13] – R. N. Datta, "Rubber curing systems", Rapra Review Report 144, Rapra Technology Ltd, 12 (12), 2002, pp 32-37. (ISSN: 0889-3144)

[1.14] – P. A. Ciullo, N. Hewitt, "The rubber formulary", Noyes publications, New York, 1999, pp 85, 91,147, and 275.

[2.1] – P. A. Ciullo, N. Hewitt, "The rubber formulary", Noyes Publications, NY 1999, p. 79.

[2.2] – http://www.westliberty.edu/health-and-safety/files/2012/08/Stearic-Acid.pdf, Date visited: 11.1.2018.

[2.3] – F. Saeed, A. Ansarifar, R. J. Ellis, Yared Haile-Meskel, "Assessing effect of the re-agglomeration and migration of chemical curatives on the mechanical properties of natural rubber vulcanizate" Adv. Polym. Technol., 32, No. S1, E153-E165 (2013).

[2.4] – R. N. Datta, "Rubber curing systems", Rapra Report 144, 12 (12) (2002), p. 2-37. ISSN:0889-3144

[2.5] – L. Bateman, C. G. Moore, M. Porter, B. Saville, "The chemistry and physics of rubber-like substances", Oxford, MRPRA (1963), 1429.

[2.6] – R. N. Datta, "Rubber curing systems", Rapra Report 144, 12 (12), 2002. ISSN: 0889-3144

[2.7] – S. H. Sheikh, X. Yin, A. Ansarifar, K. Yendall, "The potential of kaolin as a reinforcing filler for rubber composites with new sulfur cure systems". J. Rein Plas & Compos., 36 (16) (2017), 1132-1145.

[3.1] – P. A. Ciullo, N. Hewitt, "The rubber formulary", Noyes Publications, NY 1999, p. 105.

[3.2] – P. A. Ciullo, N. Hewitt, "The rubber formulary", Noyes Publications, NY 1999, p. 75.

[3.3] – B. L. Chan, D. J. Elliott, M. Holley, J. F. Smith, "The influence of curing systems on the properties of natural rubber", J. Polym. Sci., 48, (1974), p. 61-86.

[3.4] – M. Nasir, G. K. Teh, Eur. Polym. J., "The effects of various types of crosslinks on the physical properties of natural rubber", 24 (8), (1988), p. 733-736.

[3.5] – P. A. Ciullo, N. Hewitt, "The rubber formulary", Noyes Publications, NY 1999.

[3.6] – R. N. Datta, Rubber curing systems, Rapra Report 144, 12 (12), 2002

[4.1] – P. A. Ciullo, N. Hewitt, "The rubber formulary," Noyes publications, New York, 1999, p. 181.

[4.2] – P. A. Ciullo, N. Hewitt, "The rubber formulary," Noyes publications, New York, 1999, p. 281.

[4.3] – A. Ansarifar, K. Noulta, G. W. Weaver, K. G. U. Wijayantha, "Mineral kaolin for rubber reinforcement", REF Rubber Fibres Plastics, 16(4) (2021), 182.

[4.4] – A. Ansarifar, K. Noulta, S. H. Sheikh, G. W. Weaver, K. G. U. Wijayantha, "Rubber cure system", Tire Technology International, (2021), 56.

[5.1] – http://www.alibaba.com. Date visited: 15.08.2020.

[5.2] – R. N. Datta, "Rubber curing systems," Rapra Report 144, 12/12 (2002), 37.

[5.3] – R. N. Datta, "Rubber curing systems," Rapra Report 144, 12/12 (2002), 16.

[5.4] – https://nj.gov/health/eoh/rtkweb/documents/fs/2037.pdf Date visited: 17-11-2021

[5.5] – https://sustainablewayland.org/wp- content/uploads/2021/03/ZincFromCrumbRubberMays.pdf Date visited: 17-11-2021

[5.6] – https://americasinternational.com/wp-content/uploads/2018/08/TBBS-GRANULES-SDS-A.I..pdf Date visited 17-11-201

[5.7] – http://sunboss.ca/files/products/tmtd-579140.pdf Date visited 18-11-21

[5.8] – Sheth, R. N. Desai, "Replacing TMTD with nitrosamine free TBzTD accelerator in curing rubber". Inter. J. Sci. Res. Dev., 1 (3), 2013, p. 532-535.

[5.9] – https://www.fda.gov/media/147331/download Date visited 18-11-21

[5.10] – https://patents.google.com/patent/US20060137575A1/en Date visited: 19-11-2021

[5.11] – P. A. Ciullo, N. Hewitt, "The rubber formulary", Noyes Publications, New York 1999.

[5.12] – F. Saeed, A. Ansarifar, R. J. Ellis, Y. Haile-Meskel, M. Shafiq Irfan, "Two advanced styrene-butadiene/polybutadiene rubber blends filled with a silanized silica nanofiller for potential use in passenger car tire tread compound", J. Appl. Polym. Sci., 123, (2012), p. 1518-1529.

7.2 List of Abbreviations

NR: Natural rubber
SRs: Synthetic rubbers
IR: Synthetic polyisoprene
UV: Ultra violet light

BR: Polybutadiene rubber
T_g: Glass transition temperature
SBR: Styrene-butadiene rubber
EPDM: Ethylene-propylene-diene rubber
DCP: Dicyclopentadiene
ENB: Ethylidene norbornene
HX: Trans-1.4 hexadiene
NBR: Acrylonitrile-butadiene rubber
IIR: Isoprene-isobutylene rubber or butyl rubber:
CR: Poly-2-chlorobutadiene or polychloroprene
rubber
phr: parts per hundred rubber by weight
TMTD: Tetramethyl thiuram disulphide
S: sulfur
CBS: N-cyclohexyl-2-benzothiazole sulfenamide
TBBS: N-tert. Butyl-2-benzothiazole sulfenamide
MBS: 2-benzothiazole-N-sulfene morpholide
DCBS: N,N-dicyclohexyl-2-benzothiazole-
sulfenamide
5BS: tert. Amyl-2-benzothiazole sulfenamide
TBzTD: Tetrabenzylthiuram disulphide
MBT: 2-mercaptobenzothiazole
ZnO: Zinc oxide
NDPA: N-nitrosodiphenylamine
BES: N-benzoic acid
PTA: Phthalic anhydride
SCS: Salicylic acid
Semi-EV: Semi efficient vulcanization
EV: Efficient vulcanization
HMMM: Hexa-methoxymethyl-melamine
REACH: Registration, Evaluation, Authorisation,
and Restriction of Chemicals
MU: Mooney Units
ODR: Oscillating disc rheometer
CRI: Cure rate index

t_{s2}: Scorch time

t_{95}: Optimum cure time

dNm: Deci newton meter (torque)

M_L: Minimum torque

M_H: Maximum torque

ΔTorque: M_H-M_L

Min: Minutes

TRGS 522: Technical rules for dangerous substances

7.3 List of Figures and Scheme

2.7 – ΔTorque vs. ZnO loading for BR with TBBS and different amounts of sulfur. (■) 0.5 phr sulfur and 1.75 phr TBBS, (▲) 1 phr sulfur and 3 phr TBBS.

2.8 – ΔTorque vs. stearic acid loading for BR with 0.5 phr sulfur, 1.75 phr TBBS and 0.2 phr ZnO (♦), NR with 1 phr sulfur, 1.5 phr TBBS and 0.2 phr ZnO (■), EPDM with 1 phr sulfur, 1 phr TBBS and 0.075 phr ZnO (●).

2.9 – Cure traces [(torque vs. time (min)] of the BR rubber compounds with different amounts of sulfur, TBBS and ZnO. a) with 0.5 phr sulfur, 1.75 phr TBBS, and 0.2 phr ZnO; b) with 1 phr sulfur, 3 phr TBBS and 0.2 phr ZnO.

2.10 – ΔTorque vs. TBBS loading for EPDM with 1 phr sulfur.

2.11 – ΔTorque vs. ZnO loading for EPDM with 1 phr sulfur and 1 phr TBBS.

2.12 – Cure trace [(torque vs. time (min)] of the EPDM rubber compound with with 1 phr sulfur, 1 phr TBBS, and 0.075 phr ZnO.

3.1 – Typical cure traces of the rubber compounds with an increasing loading of TBBS in the powder (Table 3.1), a) Compound 1 (0.135 phr TBBS), b) Compound 6 (0.351 phr TBBS), c) Compound 10 (0.383 phr TBBS)

3.2 - ΔTorque vs. TBBS loading in the powder for the rubber compounds shown in Table 3.1.

3.3 - CRI vs. TBBS loading in the powder for the rubber compounds shown in Table 3.1.

3.4 - t_{95} and t_{s2} vs. TBBS loading in the powder for the rubber compounds shown in Table 3.1. Optimum cure time (■), scorch time (●).

3.5 - Typical cure traces (Torque (dNm) vs. Time (min)) of the rubber compounds with an increasing loading of the powder. a) compound 1 with 0.63 phr powder; b) compound 4 with 2.5 phr powder; c) compound 9 with 5.63 phr powder.

and 3.13 phr powder, f) compound with 5.5 phr sulfur and 5 phr powder.

7.4 List of Tables

8 INDEX

Symbols

A

B

C

Tetramethyl thiuram
    disulphide 10, 12, 22,
    102, 133
Tetramethyl thiuram
    monosulphide 12
Thiazole accelerator 11,
    13, 14
Thiazoles 11, 13, 14
Thiuram accelerators xii, 11,
    21, 152
Thiuram disulphides 9, 10,
    11, 12, 13, 22, 102, 133
Thiuram tetrasulphides 11
Thiuram trisulphide 12
Thiuram vulcanization 11, 12
TMTD/ZnO powder 57, 62,
    63, 64, 69, 97, 98, 99,
    100, 107, 108, 125, 126,
    127, 136, 138
Toxicity 149
Trans-1.4 hexadiene 7, 133
TRGS 522 103, 134
Triethanolamine 14, 15

U

Unsaturated hydrocarbon
    polymer 4, 5, 6, 7, 8

V

Viscosity 23, 54
Vulcanization optimum 13
Vulcanizing agents 9,
    18, 130

W

Weak amines 14
Webpage 116, 117, 118, 119,
    120, 139
Working temperature range
    4, 7

X

X/ZnO powder 127

Z

Zinc dimethyl
    dithiocarbamate 12
Zinc dithiocarbamate 12
Zinc oxide xii, xiii, 12, 19, 20,
    22, 133, 149, 150, 152
Zinc stearate 31

9 EPILOGUE

The chemicals known as accelerators and activators play a major part in the sulfur vulcanization of rubber. These chemicals and sulfur are added to the raw rubber and heated to a high temperature to produce crosslinks that give shape stability to the vulcanized rubber. Among the accelerators, the sulfenamide and thiuram types, and among the activators, zinc oxide and stearic acid, are extensively used in sulfur cure systems. It is customary to add one or two accelerators and two activators with sulfur to complete the cure. In recent years, the harmful effects of these chemicals and their adverse impacts on the environment have received a lot of attention. As a result, there are now laws and regulations that limit their excessive use because of their high level of toxicity and harmful effects on human and animal health. A new method has been developed that measures the amount of a sulfur-donor accelerator needed to provide monomolecular coverage of the zinc oxide based on the approximate surface areas of the zinc oxide and the accelerator molecule used. This new method uses only one accelerator, no stearic acid, and no sulfur to produce a single additive that offers a highly efficient cure, despite using about 80 wt% fewer chemicals in some formulations. Some studies have shown that when this additive replaces the more traditional two accelerators, two activators, and sulfur cure system in rubber formulation, the mechanical properties of the vulcanized rubber are outstanding, and in some cases, even superior to the ones reported before for the same rubber. This is a departure from the past when efficiency in sulfur vulcanization was achieved by using a high volume of these chemicals. The best way to use the accelerators is to calculate the amount needed to provide monomolecular coverage of the zinc oxide, and this is based on the approximate surface areas of the zinc oxide and the accelerator molecules used. These measurements are performed routinely. A few sulfur-donor accelerators that exist produce highly toxic nitrosamines during high-temperature curing. When safer accelerators are on hand to treat the zinc oxide, a suitable single additive will be available to replace the sulfur cure systems currently in use in industrial rubber formulations. The development of such a product will be challenging, but history has shown that no task is impossible for rubber chemists. There will be huge benefits for the manufacturers and users of rubber chemicals. It is well known that unreacted chemicals, such as zinc oxide, migrate to the surface and damage the internal structure of the rubber extensively. Moreover, chemical migration causes bloom, surface contamination, and premature rubber failure in service. This is both bad for the product and bad for the environment. This problem is solved by using fewer chemicals to cure the rubber. As a result, it is preferable to use zinc oxide to determine the precise amount of an accelerator required for a satisfactory cure rather than relying on outdated methods that need to be modified.

10 AFTERWORD

It is widely acknowledged that the rubber industry is a major contributor to the progress and well-being of mankind. Rubber is used in many applications, for example, engineering, aerospace, transportation, pharmaceuticals, adhesives, medical, electrical, electronics, seals, friction tapes, insulating materials, sports, clothes, and printing presses. The success of rubber as a useful material and the immense growth in the size of the rubber industry in the last 100 years has been truly astounding. This is mainly due to the availability of a wide range of chemicals to cure or vulcanize the rubber and solid fillers such as mineral clays, carbon black, and silica to reinforce the mechanical and dynamic properties of the cured rubber. In the past, there was little or no useful information on the detrimental effects of rubber chemicals on human and animal health and the environment. However, that is no longer the case. There is plenty of data available in technical literature describing the harmful properties of rubber chemicals, and as a result, there are now laws and regulations which limit their use. It is incumbent on rubber chemists and rubber scientists and technologists to provide an alternative route to the application of chemical curatives in the compounding of rubber. There is no need to produce more effective or more suitable additives, but rather that we should use current additives more efficiently. This may require a complete re-think of the methods used to measure them. A more efficient use of these chemicals will be highly beneficial to rubber compounders and users of rubber chemicals. This book is a collection of the experimental results and research findings from the work that tested and validated two new methods for measuring the exact amounts of zinc oxide and two sulfenamide and thiuram accelerators required for the optimum sulfur vulcanization of some industrially important rubbers. In view of the challenges that the rubber industry faces and the urgent need to limit the use of these chemicals, the publication of this book is timely. The information reported in this book will set the scene for future work to make rubber a much safer and more environmentally friendly material. There is no reason why this task cannot be achieved now that sound and efficient methods are available. In addition, a software program has been developed to help with the selection and management of zinc oxide and the accelerators used for the sulfur vulcanization of rubber.

11 AUTOBIOGRAPHY

Dr. Ali Ansarifar was awarded a bachelor's degree in Engineering, a doctorate degree in Materials Science, and a diploma in Interface Science from the University of London. He worked as a post-doctoral research assistant at Imperial College and at the Cavendish Laboratory, Department of Physics, at the University of Cambridge. He then spent several years working in a rubber research and development center in Hertfordshire, U.K., before becoming a lecturer at Loughborough University, where he retired as senior lecturer.

He has given lectures, seminars, and workshops in the United States, the United Kingdom, Europe, the Middle East, and Southeast Asia and has published over 140 technical research papers in scientific journals and in technical magazines for the polymer and tire industries. He has been on the editorial board of rubber and adhesion scientific journals and has been awarded prizes for his scientific publications. He is a Fellow of the Higher Education Academy U.K.

Dr. Ansarifar has worked in rubber research and development for the last 30 years. His expertise is in rubber-to-rubber adhesion, rubber-to-metal bonding, rubber reinforcement with solid fillers, rubber surface characterization, rubber blends, waste rubber recycling, rubber fatigue and fatigue crack growth in rubber, silica and coupling agents, chemical migration effects on the mechanical and dynamic properties of vulcanizates, and sulfur vulcanization of rubber. His main research strategy is to reduce excessive use of the chemical curatives in sulfur vulcanization and replace carbon black, silica, and silane systems in rubber reinforcement with safer and cheaper mineral clay fillers to help the rubber manufacturing industry meet its obligations under various laws and regulations for health, safety, and the environment. He is a strong advocate of a green rubber industry world-wide.

12 ACKNOWLEDGMENTS

The author acknowledges contributions made by Saad H Sheikh, Kornkanok Noulta, Dr George W Weaver, Professor K G Upul Wijayantha, and Jonathan Dushyanthan at Loughborough University, UK, and Sankha Kahagala Gamage at Bath University, UK.

The author thanks the Editorial Board of the GAK Gummi Fasern Kunststoffe and the REF Rubber Fibres Plastics International magazines of Germany for their continuous support in developing novel methods for sulfur vulcanization of rubber towards green rubber products.

The author is grateful to the Editorial Board of Tire Technology International magazine UK for their continuous support in developing novel methods for the sulfur vulcanization of rubber towards green tire compounds.

ABOUT THE BOOK

In this book, two new methods are described and supported by extensive experimental results that measure the amounts of a sulfur-donor thiuram accelerator and a sulfenamide accelerator needed to provide monomolecular coverage of zinc oxide. The methods use zinc oxide with only one accelerator and no stearic acid to produce two single additives. The additives cure rubber efficiently, despite using fewer chemicals. This is a departure from the past when efficiency in sulfur vulcanization was achieved by using a high volume of these chemicals. Although both additives are highly efficient in vulcanizing rubber, the one that was treated with the thiuram accelerator is by far more effective and can be used to vulcanize rubber without sulfur or with additional sulfur. A few sulfur-donor accelerators that exist produce highly toxic nitrosamines during high-temperature mixing and curing. When safer accelerators are on hand to treat the zinc oxide, a suitable single additive will be available to replace the current sulfur cure systems in industrial rubber formulations. This will bring huge benefits to the manufacturers and users of these chemicals. It is preferable to use zinc oxide to determine the precise amount of the accelerator required for a satisfactory cure rather than relying on outdated methods that need to be modified. A software program was also developed to help with the selection and management of zinc oxide and the accelerators for sulfur vulcanization. Green rubber products are within reach.